BENCHMARKING CURRENT PRACTICES IN PROBABILISTIC FAULT DISPLACEMENT HAZARD ANALYSIS FOR NUCLEAR INSTALLATIONS

The following States are Members of the International Atomic Energy Agency:

AFGHANISTAN
ALBANIA
ALGERIA
ANGOLA
ANTIGUA AND BARBUDA
ARGENTINA
ARMENIA
AUSTRALIA
AUSTRIA
AZERBAIJAN
BAHAMAS, THE
BAHRAIN
BANGLADESH
BARBADOS
BELARUS
BELGIUM
BELIZE
BENIN
BOLIVIA, PLURINATIONAL STATE OF
BOSNIA AND HERZEGOVINA
BOTSWANA
BRAZIL
BRUNEI DARUSSALAM
BULGARIA
BURKINA FASO
BURUNDI
CABO VERDE
CAMBODIA
CAMEROON
CANADA
CENTRAL AFRICAN REPUBLIC
CHAD
CHILE
CHINA
COLOMBIA
COMOROS
CONGO
COOK ISLANDS
COSTA RICA
CÔTE D'IVOIRE
CROATIA
CUBA
CYPRUS
CZECH REPUBLIC
DEMOCRATIC REPUBLIC OF THE CONGO
DENMARK
DJIBOUTI
DOMINICA
DOMINICAN REPUBLIC
ECUADOR
EGYPT
EL SALVADOR
ERITREA
ESTONIA
ESWATINI
ETHIOPIA
FIJI
FINLAND
FRANCE
GABON
GAMBIA, THE
GEORGIA
GERMANY
GHANA
GREECE
GRENADA
GUATEMALA
GUINEA
GUYANA
HAITI
HOLY SEE
HONDURAS
HUNGARY
ICELAND
INDIA
INDONESIA
IRAN, ISLAMIC REPUBLIC OF
IRAQ
IRELAND
ISRAEL
ITALY
JAMAICA
JAPAN
JORDAN
KAZAKHSTAN
KENYA
KOREA, REPUBLIC OF
KUWAIT
KYRGYZSTAN
LAO PEOPLE'S DEMOCRATIC REPUBLIC
LATVIA
LEBANON
LESOTHO
LIBERIA
LIBYA
LIECHTENSTEIN
LITHUANIA
LUXEMBOURG
MADAGASCAR
MALAWI
MALAYSIA
MALI
MALTA
MARSHALL ISLANDS
MAURITANIA
MAURITIUS
MEXICO
MONACO
MONGOLIA
MONTENEGRO
MOROCCO
MOZAMBIQUE
MYANMAR
NAMIBIA
NEPAL
NETHERLANDS, KINGDOM OF THE
NEW ZEALAND
NICARAGUA
NIGER
NIGERIA
NORTH MACEDONIA
NORWAY
OMAN
PAKISTAN
PALAU
PANAMA
PAPUA NEW GUINEA
PARAGUAY
PERU
PHILIPPINES
POLAND
PORTUGAL
QATAR
REPUBLIC OF MOLDOVA
ROMANIA
RUSSIAN FEDERATION
RWANDA
SAINT KITTS AND NEVIS
SAINT LUCIA
SAINT VINCENT AND THE GRENADINES
SAMOA
SAN MARINO
SAUDI ARABIA
SENEGAL
SERBIA
SEYCHELLES
SIERRA LEONE
SINGAPORE
SLOVAKIA
SLOVENIA
SOMALIA
SOUTH AFRICA
SPAIN
SRI LANKA
SUDAN
SWEDEN
SWITZERLAND
SYRIAN ARAB REPUBLIC
TAJIKISTAN
THAILAND
TOGO
TONGA
TRINIDAD AND TOBAGO
TUNISIA
TÜRKİYE
TURKMENISTAN
UGANDA
UKRAINE
UNITED ARAB EMIRATES
UNITED KINGDOM OF GREAT BRITAIN AND NORTHERN IRELAND
UNITED REPUBLIC OF TANZANIA
UNITED STATES OF AMERICA
URUGUAY
UZBEKISTAN
VANUATU
VENEZUELA, BOLIVARIAN REPUBLIC OF
VIET NAM
YEMEN
ZAMBIA
ZIMBABWE

The Agency's Statute was approved on 23 October 1956 by the Conference on the Statute of the IAEA held at United Nations Headquarters, New York; it entered into force on 29 July 1957. The Headquarters of the Agency are situated in Vienna. Its principal objective is "to accelerate and enlarge the contribution of atomic energy to peace, health and prosperity throughout the world".

IAEA-TECDOC-2092

BENCHMARKING CURRENT PRACTICES IN PROBABILISTIC FAULT DISPLACEMENT HAZARD ANALYSIS FOR NUCLEAR INSTALLATIONS

INTERNATIONAL ATOMIC ENERGY AGENCY
VIENNA, 2025

For further information on this publication, please contact:

External Events Safety Section
International Atomic Energy Agency
Vienna International Centre
PO Box 100
1400 Vienna, Austria
Email: Official.Mail@iaea.org

Printed by the IAEA in Austria
July 2025
https://doi.org/10.61092/iaea.74us-dn4n

IAEA Library Cataloguing in Publication Data

Names: International Atomic Energy Agency.
Title: Benchmarking current practices in probabilistic fault displacement hazard analysis for nuclear installations / International Atomic Energy Agency.
Description: Vienna : International Atomic Energy Agency, 2025. | Series: IAEA TECDOC series, ISSN 1011-4289 ; no. 2092 | Includes bibliographical references.
Identifiers: IAEAL 25-01765 | ISBN 978-92-0-109225-0 (paperback : alk. paper) | ISBN 978-92-0-109325-7 (pdf)
Subjects: LCSH: Nuclear facilities — Safety measures. | Nuclear facilities — Earthquake Effects. | Earthquake hazard analysis. | Hazard mitigation.

FOREWORD

IAEA Safety Standards Series No. SSR-1, Site Evaluation for Nuclear Installations, identifies fault capability as one of the potential challenges to the safety of nuclear installations. In the site selection and characterization stages of site evaluations for new nuclear installation sites, sufficient geological, geophysical, geotechnical and seismological data are obtained to demonstrate the existence of a capable fault at or near the site. Although the capable fault issues are expected to be addressed at these stages of site evaluation, subsequent studies may reveal that there is potentially a capable fault in the site vicinity of existing nuclear installations. For this case, IAEA Safety Standards Series No. SSG-9 (Rev. 1), Seismic Hazards in Site Evaluation for Nuclear Installations, recommends the assessment of the potential for fault displacement using probabilistic methods.

In the past decade, challenges associated with the application of probabilistic fault displacement hazard analysis (PFDHA) in different tectonic settings have been recognized, and new and improved empirical fault displacement models have been published. To follow up on the recent advances in this field and for a better understanding of the PFDHA results for scenarios that may be related to nuclear installations, the IAEA designed a PFDHA exercise. This publication aims to support the implementation of the recommendations given in SSG-9 (Rev. 1) by disseminating the results and findings of the IAEA PFDHA exercise. Comparison of the hazard results and the interpretation of the parts of the PFDHA that contribute to the observed differences will assist States in evaluating the suitability and performance of state of the practice fault displacement models.

The IAEA is grateful to all those who contributed to the drafting and review of this publication, in particular, S. Thompson (United States of America), E. Viallet (France) and T. Sakai (Japan) for coordinating the IAEA PFDHA exercise. The IAEA officers responsible for this publication were A. Valentini and Z. Gulerce of the Division of Nuclear Installation Safety.

CONTENTS

1. INTRODUCTION

1.1. BACKGROUND

IAEA Safety Standards Series No. SSR-1, Site Evaluation for Nuclear Installations [1] identifies fault capability as one of the potential challenges to the safety of nuclear installations. Requirement 15 of SSR-1 states:

> **"Geological faults larger than a certain size and within a certain distance of the site and that are significant to safety shall be evaluated to identify whether these faults are to be considered capable faults. For capable faults, potential challenges to the safety of the nuclear installation in terms of ground motion and/or fault displacement hazards shall be evaluated."**

Section 7 of IAEA Safety Standards Series No. SSG-9 (Rev. 1), Seismic Hazards in Site Evaluation for Nuclear Installations [2] provides recommendations to meet this requirement for new and existing nuclear installation sites. Paragraphs 7.2–7.5 of SSG-9 (Rev. 1) [2] provide recommendations on defining the conditions for fault capability, and principal and distributed fault displacements; recommendations on the necessary investigations to determine fault capability are provided in paras 7.6–7.9 of SSG-9 (Rev. 1) [2]. Based on these definitions and the results of the site investigations, SSG-9 (Rev.1) [2] provides recommendations on methods and practices to be used with regard to the evaluation of capable faults for proposed new sites (paras 7.10 and 7.11) and for sites with existing nuclear installations (paras 7.12–7.18).

In the site selection and characterization stages of site evaluations for new nuclear installation sites, sufficient geological, geophysical, geotechnical, and seismological data are expected to be obtained to demonstrate the existence of fault capability at or near the site. Although the capable fault issues are expected to be addressed in the site evaluation, subsequent studies may reveal that there is potentially a capable fault in the site vicinity of an existing nuclear installation. In such cases, paras 7.14 and 7.15 of SSG-9 (Rev. 1) [2] recommend the assessment of the potential for fault displacement, based on the slip characteristics and geometry of the fault, using probabilistic methods. The methods for developing the primary and distributed fault displacement hazard curves are generally known as probabilistic fault displacement hazard analysis (PFDHA), for which different methodologies exist in the current literature, depending on the availability of relevant data.

Reference [3] provides the most relevant application examples of PFDHA for nuclear installation sites and presents two case studies (for Diablo Canyon and Krško 2 Nuclear Power Plants (NPPs)) that used the earthquake approach and the case of the Yucca Mountain Geological Repository that included both the displacement and earthquake approaches. These examples highlight the considerable uncertainties involved in estimating the fault displacement hazard and underline that there is limited experience in nuclear projects for performing PFDHA. In the last decade, a better understanding of the challenges associated with the application of PFDHA for different projects in different tectonic settings have been developed. More comprehensive datasets of surface ruptures were compiled [4, 5] and new and improved empirical models with better parametrization in terms of slip and site specific geology were published (e.g. Ref. [6]).

1.2. OBJECTIVE

To follow up with the recent advances in this field and for a better understanding of the PFDHA results for scenarios that may be related to nuclear installations, a PFDHA exercise was designed by IAEA. This exercise was focused on the principal and distributed fault displacements estimated for earthquakes with different style of faulting. Two earthquakes occurred in 2016: the magnitude 7.0 Kumamoto earthquake, which represents strike-slip faults; and the magnitude 6.5 Norcia earthquake associated with normal faulting. These were selected as representative cases, along with the magnitude 4.9 reverse Le Teil earthquake that occurred in 2019. Alternative seismic source characterization models were developed for each case and all seismic source model parameters, including the site coordinates and dimensions, were provided to the analysts to the ensure consistency among PFDHA results of different modelling teams.

The main objective of this publication is to disseminate the results and findings of the IAEA PFDHA exercise. Comparison of the hazard results and the interpretation of the parts of the PFDHA that contribute to the observed differences will support States in evaluating the suitability and performance of state of the practice fault displacement models. The detailed technical elements of PFDHA for scenarios that are relevant for nuclear installation sites are discussed, to assist in the implementation of the recommendations provided in SSG-9 (Rev. 1) [2].

1.3. SCOPE

The scope of this publication is the IAEA PFDHA exercise, which was designed and parameterized by considering the fault capability constraints for nuclear installation sites. The lessons learned during this interactive exercise may be implemented in the review and assessment of fault displacement hazards for existing NPPs and other nuclear installations in any seismotectonic environment.

This publication provides a reference text for experienced users regarding recent developments in PFDHA. It is intended for use by regulatory bodies responsible for establishing regulatory requirements and by operating organizations directly responsible for the evaluation of seismic and fault displacements hazards at a nuclear installation site.

1.4. STRUCTURE

Section 2 provides the definitions for principal and distributed fault displacement, in order to clarify the differences between the terminology used in IAEA publications and recent fault displacement prediction models. Section 3 presents the necessary technical background for PFDHA, including the surface rupture models, fault displacement prediction models and equations used in the IAEA PFDHA exercise. Section 4 describes the general framework of the IAEA PFDHA exercise and introduces different groups that participated to the study by providing their models or by performing the hazard calculations. Sections 5–7 present the case studies, source model parameters and hazard results for the Kumamoto, Le Teil and Norcia earthquakes, respectively. Section 8 presents general conclusions and discusses the lessons learned from the exercise. Section 9 seeks to develop the connection between the fault displacement analysis results and the complementary engineering analysis, which would be useful for States for external events safety assessment. The Annex builds the connection with the fault capability analysis and PFDHA, particularly by explaining the selection of the time frame for concerns regarding fault capability.

2. DEFINITIONS USED IN FAULT DISPLACEMENT HAZARD ANALYSIS FOR NUCLEAR INSTALLATIONS

This section describes the definitions and supporting information for the key terminology used in fault displacement hazard assessment. While some definitions have been described in preceding IAEA publications (e.g. Ref. [3]), recent advancements in data collection methods and the development of new predictive models for fault displacement warrant further clarification of definitions and terminology.

Relevant definitions described in previous IAEA publications are identified and discussed in Section 2.1. A brief description of fault displacement characteristics and data is given in Section 2.2 to provide context for new terms and their definitions used in hazard models. These new terms and definitions used in fault displacement models are described in Section 2.3.

2.1. DEFINITIONS USED IN PREVIOUS IAEA PUBLICATIONS

SSG-9 (Rev. 1) [2] describes the evaluation of potential fault displacement at a site and includes definitions of fault displacement, tectonic features that can cause fault displacement, and the difference between principal faults and distributed faults. In particular, para. 7.2 of SSG-9 (Rev. 1) [2] states:

> "Fault displacement is the relative movement of two sides of a fault at or near the surface, measured in any chosen direction, generated by an earthquake. Primary, or principal, faulting occurs along a main fault rupture plane (or planes) that is the location of release of the energy. Secondary, or distributed, faulting is the rupture that occurs near the principal faulting, possibly on splays of the main fault or on antithetic faults. In other words, displacements could be associated with the causative (i.e. seismogenic) fault or could occur coseismically on secondary faults. Tectonic relative displacements associated with folds (i.e. synclines and anticlines) are also included in the term 'fault displacement'. Fault creep, when demonstrated as such, is considered as a slowly progressing geological hazard that might affect the safety of the nuclear installation but is not seismically induced".

These definitions of fault displacement and responsible tectonic features are also adopted throughout this publication.

Paragraph 7.16 of SSG-9 (Rev. 1) [2] describes the definitions of faulting, and the relative movement with respect to the displacement vector, and states (footnote removed):

> "In the probabilistic fault displacement hazard analysis, the following two types of possible displacement should be considered, with careful and appropriate treatment of the uncertainties involved (both epistemic and aleatory):
>
> (a) Primary or principal displacement that occurs along a main plane (or planes) that is the locus of release of seismic energy.
> (b) Secondary or distributed displacement that occurs in the vicinity of the principal displacement, possibly on splays of the main fault or antithetic faults. In some cases, triggered slip has been considered a form of secondary or distributed displacement.
> (c) The fault displacement is generally characterized as a three-dimensional displacement vector that is resolved into components of slip along the fault trace

and along the fault dip, with the resulting amplitude equal to the total evaluated slip (for a given annual frequency of exceedance and for a given fractile of hazard)."

These definitions — using the terms 'principal' and 'distributed' rather than 'primary' and 'secondary' — are used throughout this publication.

2.2. FAULT DISPLACEMENT CHARACTERISTICS AND DATA

Although the definitions of fault displacement, responsible tectonic features, and principal and distributed faulting are described in Section 2.1, further discussion and clarification is needed to describe how fault displacement occurs, how it is measured, and how the data are used in empirical fault displacement models.

2.2.1. Fault displacement characteristics

The pattern of fault displacement at the ground surface can vary along the length of a rupture. Fault displacement may be restricted to a single narrow zone, or it may be spatially distributed across a broad and complex network of discrete faults and fractures. In both instances, the fault displacement occurs as offsets on localized, discrete faults. Other surface deformations (e.g. warping, tilting) can accompany discrete fault ruptures and produce displacement across an area with little or no localization due to non-brittle deformation. This is referred to as flexural deformation in Ref. [3].

Displacement that is measured across a single rupture, is the discrete fault displacement that occurred on that rupture plane. The width of this discrete displacement (and therefore also the measurement aperture) may be anywhere between a few metres to less than several centimetres. Cumulative fault displacement measurements across zones capture displacement from non-localized deformation and displacement from any discrete ruptures in the zone (Fig. 1). This is sometimes referred to as 'total' fault displacement [7, 8]. It is notable that 'total' in this instance is unrelated to the displacement vector component.

On-fault displacement generally refers to principal fault displacement, but off-fault displacement is inconsistently used in scientific literature [3]. The term off-fault displacement can describe discrete fault displacement on distributed faults, or non-localized surface deformations across wide apertures.

2.2.2. Fault displacement data

Several methods are used to document surface rupture and measure fault displacement; Ref. [3] presents an overview of the different methods. A brief summary of the methods and the type of data they provide, with respect to the definitions of fault displacement used in this publication, is provided below.

Most traditional field mapping datasets provide measurements of discrete fault displacement. In traditional mapping methods, surface ruptures are usually identified by geologists and located on topographic base maps or low-altitude aerial photographs. Fault displacements are typically measured in the field across discrete ruptures using piercing points (Fig. 1) with tools such as measuring tapes, folding rulers, or levelling staffs. The measurement aperture is rarely reported and is usually assumed to be less than 10 metres.

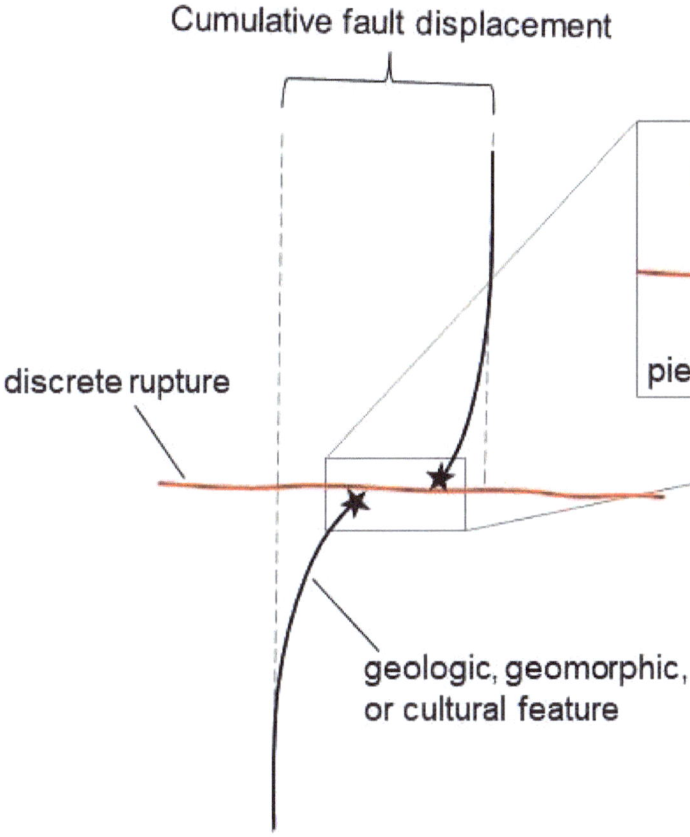

FIG.1. Schematic plan view (not to scale) showing right-lateral offset of originally linear feature. Discrete fault displacement is measured from offset piercing points on opposite sides of the fault (inset). Cumulative fault displacement includes both discrete and not localized surface deformations (main) (reproduced from Ref. [5] with permission).

Some field-based methods can capture cumulative fault displacement. For example, tools such as hand levels, levelling staffs, and surveying instruments are used to create scarp profiles and baseline surveys that provide wider aperture measurements (Fig. 1). These types of measurement will capture surface deformations that are not localized, in addition to any discrete fault displacement that may be present.

Remote sensing methods may be used to indicate cumulative fault displacement, discrete fault displacements, or both. More traditional remote sensing methods measure offset features observed in imagery such as aerial photographs from aircraft, satellites, or drones, or digital elevation models from photogrammetry or lidar. Emerging methods utilize pre- and post-earthquake data, such as lidar or orthorectified photography, and use various types of change detection or differencing methods to calculate displacement at pixels (in imagery) or points (in lidar) in the dataset [8–10]. The cumulative fault displacement is measured across a zone from the pixel or point displacements. Fault displacement can also be resolved with interferometric synthetic aperture radar when multiple interferograms (e.g. ascending and descending track pairs) are used [11, 12]. Except for very high-resolution datasets, the measurement aperture in all remote sensing methods is wider than any discrete rupture and therefore captures some amount of non-localized surface deformation, such as warping or tilting.

Most earthquake datasets only contain measurements for the dominant component associated with the style of faulting. For example, datasets for dip-slip events commonly only report vertical displacement measurements, and strike-slip event datasets often only report lateral displacement. Datasets with complete measurements for the lateral and vertical components of the displacement vector are rare, and measurements that include the fault-perpendicular horizontal component (e.g. shortening or extension; also called heave) are extremely rare. Furthermore, datasets rarely distinguish between whether the minor displacement components were observed to be zero or were simply not measured, such as the lateral component in dominantly dip-slip (i.e. reverse, or normal) events.

Some datasets classify surface ruptures and related fault displacement measurements as principal or distributed; however, this classification is more commonly performed as part of the dataset compilations that are specific to fault displacement hazard analysis model development ([4, 5, 13–15]). Although the definitions of principal and distributed faulting in SSG-9 (Rev. 1) [2] are common within the fault displacement hazard community, the process of assigning the classifications is ultimately subjective and non-unique, especially for complex surface ruptures. Furthermore, the assignment of a rupture trace or displacement as principal or distributed is based on the observations and measurements collected following the earthquake, with little documentation of whether the rupture trace was or could have been mapped before the earthquake as a principal or distributed fault [16, 17].

2.3. DEFINITIONS USED IN FAULT DISPLACEMENT HAZARD ANALYSIS

Both the probability of surface rupture models and the models that predict the amount of fault displacement are used in fault displacement hazard analysis.

2.3.1. Surface rupture definition

Surface rupture is not explicitly defined in IAEA publications [2, 3]. It is typically defined as co-seismic fault displacement that breaches the ground surface. This includes discrete planar scarps that abruptly break the ground surface and more complex scarps that involve broken soil or rock [18, 19]. Fault-related folds are also usually considered surface rupture, and para. 7.2 in SSG-9 (Rev. 1) [2] specifically considers fault displacements associated with folds. Throughout this publication, surface rupture is considered to include scarps, mole tracks, and fault-related folding.

Models that predict the likelihood of surface rupture occurrence are called 'probability of surface rupture models'. The models may be purely statistical or derived from numerical simulations (Section 3.1). Different models are available for the probability that surface rupture occurs anywhere in the earthquake (e.g. Refs [20, 21]) and at the site location (e.g. Ref. [22]).

2.3.2. Fault displacement definition

Models that predict the amount of fault displacement are called fault displacement models. The definitions of fault displacement used in SSG-9 (Rev. 1) [2] are broadly consistent with the definitions used in empirical fault displacement models. However, specific details can vary between different models due to differences in the datasets and modelling assumptions. More specific definitions are needed to correctly apply model results to engineering analyses, compare results between models, and appropriately use alternative models to capture epistemic uncertainty. Fault displacement is therefore more completely defined in fault displacement models with respect to the displacement vector and whether the measurement represents only discrete displacement or captures some amount of non-localized surface deformation, such as warping or tilting. This does not replace the definitions used in SSG-9 (Rev. 1) [2], but rather expands on them.

Surface ruptures produce three-dimensional ground surface displacement, and the displacement vector is the three-dimensional component of displacement as described in paras 7.2 and 7.16 of SSG-9 (Rev. 1) [2]. The full vector is the net displacement, and it consists of three orthogonal components: fault-parallel horizontal (i.e. lateral), fault-perpendicular horizontal (i.e. heave, or shortening/extension), and vertical (i.e. throw). Components of this vector are measured (in data) and/or predicted (in models). However, most datasets do not provide all the three components nor a measurement of the net vector. This incompleteness is an inherited limitation in fault displacement models. As a result, practitioners need to understand if a component is unintentionally not included in a fault displacement model and consider that it may need to be estimated by other means.

Most fault displacement models are developed for either principal or distributed faults (e.g. Refs [13, 14, 22–25]). The currently available fault displacement models address discrete fault displacement. While some models included data that represent cumulative fault displacement across relatively narrow apertures (e.g. the vertical displacement from scarp profiles across broad folds in reverse-faulting earthquakes), these data were treated as discrete fault displacement in the model development. Established methods for estimating the amount and spatial distribution of non-localized surface deformation are not currently available.

The predictions in most models are usually understood to be for a single discrete rupture, although this might not be explicitly stated in the data or models. As a result, the effect of surface rupture complexity on the amount of fault displacement is not captured in most models. Figure 2 shows an example of a complex surface rupture pattern with (sub)parallel principal ruptures. Principal fault displacement measurements on different ruptures are projected onto a centreline, effectively treating measurements on different faults as adjacent along a single nominal strike line.

The available fault displacement models cannot predict aspects of fault complexity, such as the number of fault ruptures, nor the amount of discrete displacement on individual ruptures. As an alternative, some new models provide predictions for the combined discrete displacement on all ruptures (e.g. Refs [26–28]) by performing an additional data processing step to combine (sum) discrete fault displacement measurements on (sub)parallel ruptures across a zone as demonstrated in Fig. 2. Displacement from distributed faults may or may not be included. When displacements from only principal ruptures are considered, this is referred to as the sum of principal displacement [6]; when distributed displacements are included, this is called aggregate displacement [6, 26, 28]. Practitioners need to consider how to make meaningful comparisons between fault displacement models that predict single rupture principal displacement, summed principal displacement, and aggregate displacement.

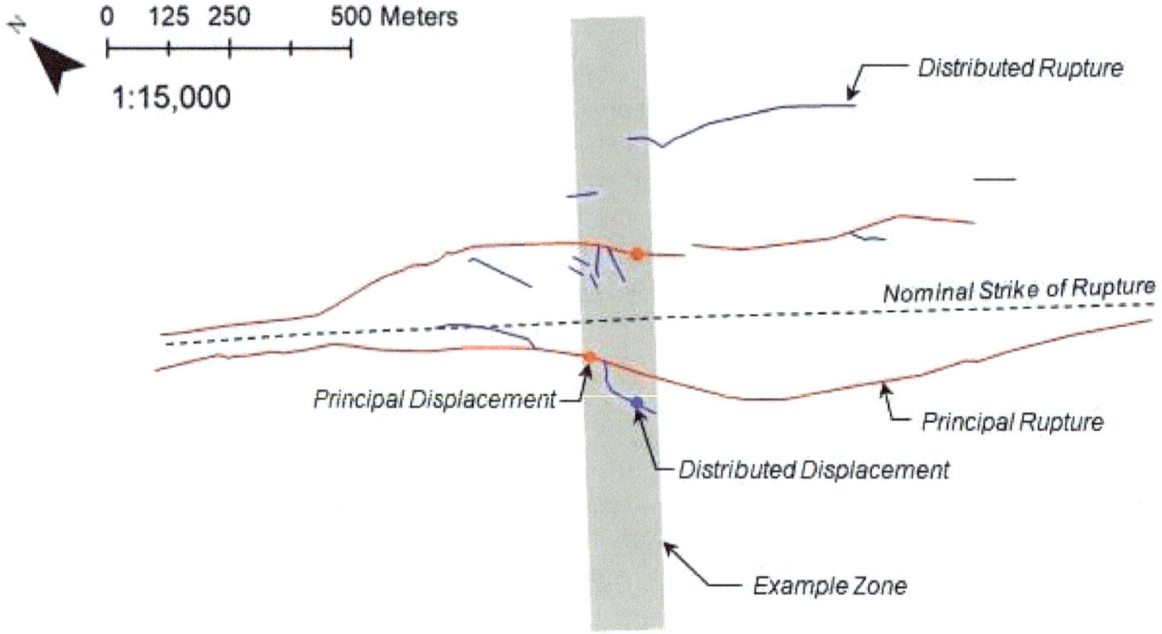

FIG. 2. Example map of surface ruptures and fault displacement measurement sites (reproduced from Ref. [6] with permission).

3. OVERVIEW OF PROBABILISTIC FAULT DISPLACEMENT HAZARD ANALYSIS AND AVAILABLE MODELS

Probabilistic fault displacement hazard analysis was first introduced by Ref. [13] and two alternative approaches called the displacement approach and earthquake approach, were established in the last two decades. A comprehensive summary of the displacement and earthquake approaches in PFDHA is provided in Ref. [3]. The IAEA PFDHA exercise focused on the earthquake approach. Accordingly, the fault displacement models, and probability of surface rupture models described in this publication are based on the earthquake approach. These models are used to estimate the annual rate of exceedance of co-seismic surface fault displacement at a site using the PFDHA integral, as given in Eq. (1):

$$\lambda(D > D_0) = \alpha(M_{min}) \int_{m,s} f_{M,S}(m,s) \times P[sr \neq 0|m] \times \int_{r} f_R(r) \times P[D \neq 0|z,r,sr \neq 0] \times P[D > D_0|l^*, m, D \neq 0] \times dr\, dm\, ds \quad (1)$$

where D is displacement at a site of area z^2 located at location l^* relative to the rupture. The state of practice distinguishes principal displacement using capital D from distributed displacement using lower case d.

The first term in Eq. (1) is the activity rate of the fault, $\alpha(M_{min})$, which is the cumulative rate of earthquakes with magnitudes larger than a given minimum magnitude (M_{min}). The outer integral in Eq. (1) integrates over the probability density function $f_{M,S}(m,s)$, which describes the joint probability of a magnitude m earthquake occurring along the fault at distance s from the end of the fault, as shown in Fig. 3(A). This notation is not a strict expression of how to numerically evaluate the integral, but rather is an abbreviated notation from Ref. [14]. It is meant to indicate that ruptures for magnitude m are to be floated along the length of the fault. Floating the earthquakes along the fault is also common practice in probabilistic seismic hazard assessment (PSHA).

The next term, $P[sr \neq 0|m]$, is the probability that an earthquake with magnitude m produces surface rupture (*sr*), anywhere along the fault. This term is unique to PFDHA, and it is needed because not all earthquakes produce surface rupture. Empirical evidence shows a strong magnitude dependence on the probability of an earthquake rupturing the surface, although other predictor variables may affect this probability (e.g. the style of faulting).

The inner integral in Eq. (1) evaluates the probability density function $f_R(r)$, which describes the aleatory variability in the location of the mapped fault system (Fig. 3(B)). There may be aleatory variability in the location as the earthquakes do not re-rupture the exact same surface fault in every earthquake. Additionally, epistemic uncertainty due to mapping uncertainties are difficult to separate from aleatory variability and may be combined in the $f_R(r)$ term [14]. Figure 3 illustrates the scenarios where the site is located on the mapped fault system (Fig. 3(A)) and off the mapped fault system but within the range of the variability of the mapped fault location (Fig. 3(B)).

The $P[D \neq 0|z, r, sr \neq 0\,]$ term is the probability of having non-zero displacement at fault-normal distance r from the source, within a given site area z^2, provided that the surface rupture (*sr*) has occurred in the earthquake. In other words, this term effectively provides the 'conditional probability of surface rupture at the specific site'. This term is needed because even when an earthquake produces surface rupture, the fault rupture might not pass through the site of interest. The dependence of this probability on r and z calls for a computational effort to calculate the probability of non-zero displacement considering the area of interest, its distance

r from the source, and the aleatory variability in the distance r. This is discussed in more detail in section 3.3.2 of Ref. [3] and in Refs [14, 29]. More explicit definitions for the source-to-site distance r, such as the location with respect to the footprint of the structure (Fig. 3(C)) or beyond the endpoints of a mapped principal fault (Fig. 3(D)), are not acknowledged in the fault displacement models. However, these issues may be important for nuclear installation sites. The conditioning parameters in $P[D \neq 0|z, r, sr \neq 0\,]$ are typical for the current practice but may change if new models are developed with different predictor variables (e.g. l^*).

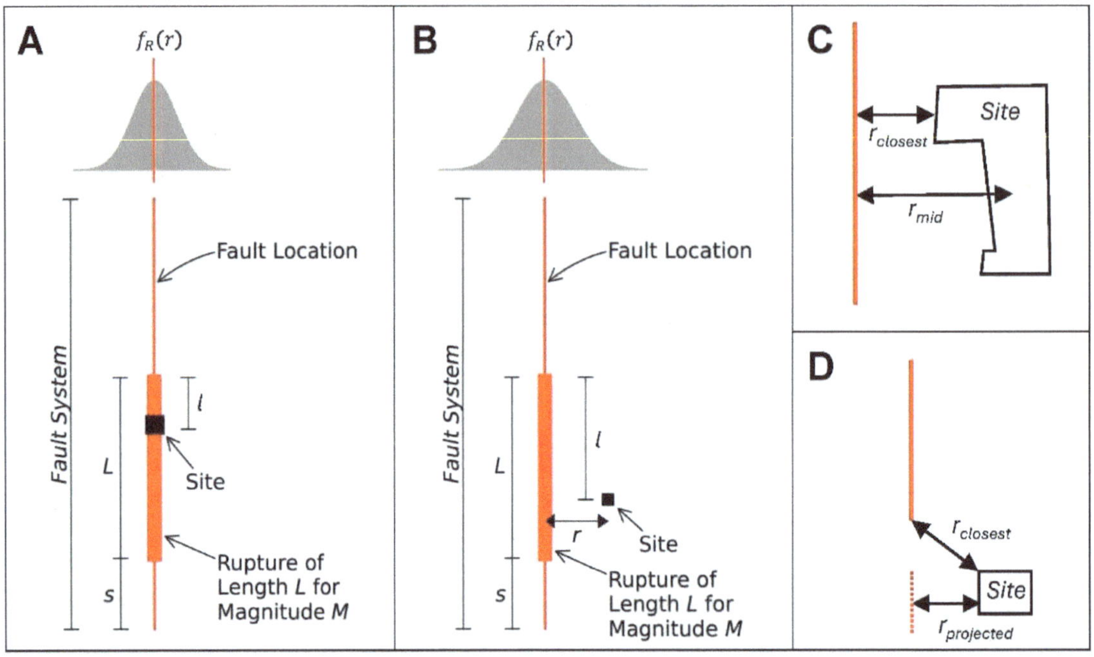

FIG.3. Illustration showing how the rupture of length L for magnitude m earthquake is advanced or floated along the fault system by incrementally increasing s. Aleatory variability in the mapped fault location is represented with $f_R(r)$. (A) Example where site is located on the mapped fault. (B) Example where site is not located on the mapped fault. (C) Alternative definitions of site-to-rupture distance r based on structural footprint. (D) Alternative definitions of site-to-rupture distance r based on end-of-rupture location.

The last term ($P[D > D_0|l^*, M, D \neq 0\,]$) is the probability that the displacement D exceeds a test level D_0, given the location parameter l^*, magnitude m, and the condition that rupture occurs at the site causing non-zero displacement ($D \neq 0$). This probability of exceedance is calculated using fault displacement models. In current fault displacement models, the location parameter l^* differs based on whether the displacement is principal or distributed.

Models that address the $P[sr \neq 0|m]$ and $P[D \neq 0|z, r, sr \neq 0\,]$ terms are called probability of surface rupture models because they are related to the likelihood of surface rupture occurrence (or zero displacement) but do not predict the amount of fault displacement. Models that describe the displacement amplitude are called fault displacement models, and they provide displacement as a probability distribution from which displacement exceedance is calculated.

It is implicitly assumed in the state-of-practice that style of faulting (i.e. fault mechanism) is also a conditioning term in the probability of surface rupture and fault displacement models. For example, Ref. [21] provides different probability of surface rupture models based on style of faulting, and the available fault displacement models are developed for specific styles.

There is an important distinction between models applicable to the principal fault displacement hazard, distributed fault displacement hazard, or both. Equation (1) is modified into Eq. (2) for probabilistic principal fault displacement hazard analysis:

$$\lambda(D > D_0) = \alpha(M_{min})\int_{m,s} f_{M,S}(m,s) \times P[sr \neq 0|m] \times \int_r f_R(r) \times P[D \neq 0|z,r,sr \neq 0] \times P\left[D > D_0|\frac{l}{L}, m, D \neq 0\right] \times dr\,dm\,ds \quad (2)$$

where D and D_0 explicitly represent principal displacement. Also, the location parameter l^* in Eq. (1) is replaced with the along-rupture distance ratio, l/L, which is the location parameter used in current fault displacement models for principal displacement (Fig. 3(A)). L is the total surface rupture length on the fault source, and l is the distance from the nearest point on the principal fault rupture to the closest end of the rupture.

The formulation for distributed fault displacement hazard is provided in Eq. (3):

$$\lambda(d > d_0) = \alpha(M_{min})\int_{m,s} f_{M,S}(m,s) \times P[sr \neq 0|m] \times \int_r f_R(r) \times P[d \neq 0|z,r,sr \neq 0] \times P[d > d_0|r, m, d \neq 0] \times dr\,dm\,ds \quad (3)$$

where d and d_0 explicitly represent distributed displacement, and the location parameter of interest is r, which is the closest distance between the principal rupture and the site of interest (Fig. 3(B)). The conditioning parameters in Eq. (3) are typical for current practice and may change if new models are developed with different predictor variables. It is notable that the probability of surface rupture ($P[sr \neq 0|m]$) is used in both Eqs (2) and (3).

Following the pioneering work of Ref. [13], several authors have proposed different predictive models for the various elements of the hazard integrals given in Eqs (2) and (3). Section 3.1 summarizes available models for the probability of surface rupture, which is common to both principal and distributed faulting. Section 3.2 summarizes models for principal fault displacement hazards, and Section 3.3 summarizes models for distributed fault displacement hazards. The IAEA PFDHA exercise uses several of these models, and the discussions in this section are intended to help the reader to understand the models and associated limitations. Section 3.4 demonstrates how fault displacement models and probability of surface rupture models affect the hazard curves. It is provided to assist readers of this publication in understanding the results from different hazard analysts given in Sections 5–7.

3.1. MODELS FOR PROBABILITY OF SURFACE RUPTURE

The conditional probability that an earthquake ruptures the ground surface at any location is represented by the $P[sr \neq 0|m]$ term in Eqs (2) and (3). Examples of empirical models to calculate the probability of surface rupture are listed in Table 1. This table is not an exhaustive list of empirical surface rupture models. All models in Table 1 use logistic regression with earthquake magnitude as the predictor variable. Models differ based on the subset of empirical data used in the regression, with most models developed for a specific style of faulting, geographic region, inferred near-surface soil condition, or a combination of these factors.

TABLE 1. PROBABILITY OF SURFACE RUPTURE MODELS

Reference	Style of Faulting*	Notes	Database	Applicable Magnitude Range	Used in This Study
[20]	All		276 global earthquakes	5.0–8.2	Yes
[13]	N		105 extensional Cordillera earthquakes	4.5–7.6	Yes
[13]	N		32 Great Basin earthquakes	4.9–7.2	Yes
[13]	N		47 northern Basin & Range earthquakes	4.9–7.4	Yes
[23]	R		129 global earthquakes	5.5–8.0	Yes
[22]	R and SS		107 Japanese earthquakes	5.5–7.4	Yes
[30]	R and SS	Models for stiff and soft soil	126 (R) and 140 (SS) global earthquakes	4.2–8.7	Yes
[31]	R		11 Australian earthquakes	4.7–6.6	No
[21]	All		363 global earthquakes	5.5–7.9	No
[21]	N		60 global earthquakes	5.5–7.1	No
[21]	R		119 global earthquakes	5.5–7.9	No
[21]	SS		139 global earthquakes	5.5–7.8	No

*N: Normal, SS: Strike-slip, R: Reverse

Figure 4 compares the probability of surface rupture for the models in Table 1 that were used in the PFDHA exercise. The plots show empirical models that may be considered for specific styles of faulting as well as the model used in Ref. [20] that is based on global data of all styles of faulting. Surface rupture probability increases with magnitude for all models, but the rate of increase with magnitude and the absolute probability at any magnitude level is highly dependent on regionalization and style of faulting. For example, the three alternative models for normal faulting yield probabilities ranging from 27% to 52% for M_W=6.0 (M_W, moment magnitude), suggesting that the geographic location of collected data (regionalization) is important. Compared to normal and strike-slip cases, the probability of surface rupture for reverse faulting is generally lower. For example, the models used in Refs [23, 30] for soft soil conditions yield probabilities of less than 50% for M_W=7.0, whereas for other styles of faulting the probability at this magnitude exceeds 70%.

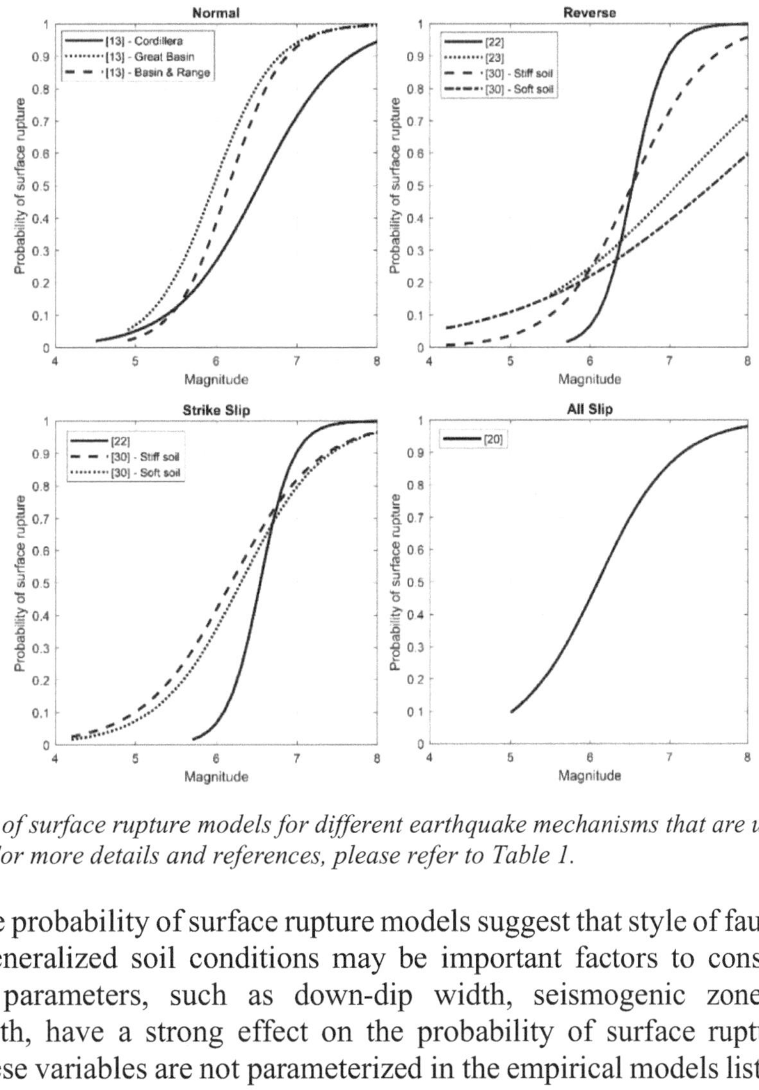

FIG. 4. Probability of surface rupture models for different earthquake mechanisms that are used in the IAEA PFDHA exercise. For more details and references, please refer to Table 1.

Evaluation of the probability of surface rupture models suggest that style of faulting, geographic location, and generalized soil conditions may be important factors to consider in selecting models. Other parameters, such as down-dip width, seismogenic zone thickness, and hypocentral depth, have a strong effect on the probability of surface rupture in numerical simulations. These variables are not parameterized in the empirical models listed in Table 1. As discussed in Section 3.3, the probability of surface rupture can have a significant impact on the hazard, so practitioners need to consider if empirical models are applicable to their site. Case-specific numerical estimations for surface rupture probability may be particularly important to consider if a specific fault or tectonic environment is different than the data used to develop the empirical models.

3.2. PRINCIPAL FAULT DISPLACEMENT

PFDHA for principal faulting is calculated using Eq. (2). Available models for the conditional probability of exceedance (i.e. fault displacement models) and conditional probability of surface rupture for principal faulting are discussed in Sections 3.2.1 and 3.2.2, respectively. At the time of this publication, very few alternative models were available for the conditional probability of surface rupture for principal faulting.

3.2.1. Conditional probability of exceedance on the principal fault

The conditional probability of exceedance is represented by the $P[D > D_0 \,|\, l/L, m, D \neq 0\,]$ term in Eq. (2). In this term, the probability that the principal displacement will exceed D_0 at a given location l/L along the surface rupture for a given magnitude M earthquake is calculated using principal fault displacement models. These models provide a probability density function for the amount of principal displacement, from which the probability of exceedance is readily calculated. Available models in the literature use different statistical distributions (e.g. normal, gamma, beta) and data transformations (e.g. logarithmic, power-normal, Box-Cox). It is notable that this practice differs from that in ground motion modelling, in which most models use a lognormal distribution.

Tables 2 and 3 list principal fault displacement models. Well-known models from Refs [13, 14, 22, 30] are included in these tables. These lists might not be exhaustive, and practitioners are encouraged to perform a literature review to compile and consider all potentially relevant models for their site. Models that predict a normalized displacement (i.e. a displacement normalized by the predicted maximum or average displacement) are listed in Table 2, whereas models that directly predict displacement amplitude are given in Table 3. The difference between these two classes of models is discussed below.

A set of four new fault displacement models was published in 2022–2024 through the Fault Displacement Hazard Initiative (FDHI) project coordinated by the University of California, Los Angeles. The new models are also listed in Tables 2 and 3. In some cases, the FDHI models supersede previously published models. For example, the new model proposed by Ref. [33] is an update to the model given in Ref. [23], and the new model by proposed by Ref. [27] is an update to the model given in Ref. [14]. The key differences between the new FDHI models and the earlier models [13, 14, 22, 23] are related to database size, statistical modelling methods, and the definition of displacement. The FDHI database contains more events for each style of faulting than previous compilations [5]. The FDHI models use advanced statistical modelling approaches such as magnitude scaling breakpoints, separation of magnitude- and location-dependent aleatory variability, event-specific effects to capture within- and between-event variability, and compute within-model epistemic uncertainty. Most of the FDHI models predict an 'aggregate' or 'summed principal' displacement (see Section 2.3). A summary and comparison of the FDHI fault displacement models are provided in Ref. [6].

The currently available fault displacement models only capture the effects of earthquake size (M) and location along the rupture (l/L), as reflected by the conditional probability of exceedance term in Eq. (2). The models are developed using data for a specific style of faulting; therefore, the effects of fault kinematics are captured by the data used in model development.

Two approaches are used to capture the effects of earthquake size. The first approach uses a normalized displacement as the predicted variable in the model formulation (Table 2). These models predict displacement as a ratio normalized by the maximum or average displacement (i.e. *D/MD* or *D/AD*). To determine *D*, a separate scaling relation is used to estimate *MD* or *AD*. The suggested approach is to predict *MD* or *AD* as a function of earthquake magnitude using empirical relations [34], where the epistemic uncertainty on *MD* or *AD* is included in the logic tree (see Section 8). In the second approach, *D* (or a transformed version, such as *ln(D)*) is directly predicted, and the earthquake magnitude is a predictor (input) variable in the model (Table 3).

Figure 5 presents displacement profiles for the normalized displacement models in Table 2 that are used in the PFDHA exercise for 50% non-exceedance probability (see last column of Table 2). These plots show how the predicted median displacement varies as a function of normalized position along the rupture length (*l/L*). All the models presented in Fig. 5 use functional forms that are symmetrical about the rupture midpoint (*l/L=0.5*). In most cases, the displacement amplitude tapers at rupture endpoints (i.e. closer to *l/L=0*) and is highest at the midpoint; the exceptions are location-independent models. Differences between slip profiles are due to a combination of different data sets used to develop the models and different modelling decisions on the shape or functional form of the displacement profile. For example, Ref. [14] provides three alternative profile shapes: bilinear, quadratic, and elliptical.

Similar plots are presented in Fig. 6 for the displacement models listed in Table 3 that are used in the PFDHA exercise for 50% non-exceedance probability (see last column of Table 3). The profiles are shown for a M_W=6.5 earthquake. For most models, the profile shape is independent of earthquake size (i.e. self-similar scaling); however, the profile shape changes with earthquake size in the models proposed by Refs [26, 33].

As discussed in Section 2.2.2, different definitions of displacement are used in the models listed in Tables 2 and 3. Table 4 summarizes the definitions of displacement used in the fault displacement models.

TABLE 2. PRINCIPAL FAULT DISPLACEMENT MODELS THAT PREDICT NORMALIZED DISPLACEMENT AMPLITUDE.

Reference	Style of Faulting	Predicted Variable	Statistical Distribution	Predictor Variables	Notes	Database	Applicable Magnitude Range	Used in This Study
[13]	N	*D/MD*	Beta	*l/L*	$0 \leq l/L \leq 0.5$	5 Basin and Range events from Ref. [35]	Not reported	No
[13]	N	*D/MD*	Beta	*l/L*	$0 \leq l/L \leq 0.5$	11 global events from Ref. [36]	Not reported	Yes
[13]	N	*D/AD*	Gamma	*l/L*	$0 \leq l/L \leq 0.5$	(See above)	Not reported	Yes
[23]	R	*D/MD*	Beta	*l/L*	$0 \leq l/L \leq 0.5$	8 global events from Ref. [37], 1 event from Ref. [23]	5.5–8.0	Yes
[23]	R	*D/AD*	Gamma	*l/L*	$0 \leq l/L \leq 0.5$	(See above)	5.5–8.0	Yes
[23]	R	*D/AD*	Weibull	*l/L*	$0 \leq l/L \leq 0.5$	(See above)	5.5–8.0	No
[14]	SS	*ln(D/AD)*	Normal	*l/L*	bilinear profile $0 \leq l/L \leq 0.5$	13 global events from Ref. [23] 9 global events from Ref. [14]	6.0–8.0	No
[14]	SS	*ln(D/AD)*	Normal	*l/L*	quadratic profile $0 \leq l/L \leq 0.5$	(See above)	6.0–8.0	No
[14]	SS	*ln(D/AD)*	Normal	*l/L*	elliptical profile	(See above)	6.0–8.0	No
[22]	R and SS	*D/MD*	Beta	*l/L*	$0 \leq l/L \leq 0.5$ for $L \geq 10$ *km*	17*	5.8–7.4*	No
[22]	R and SS	*D/AD*	Gamma	*l/L*	(See above)	(See above)	(See above)	Yes
[22]	R and SS	*D/MD*	Beta	None	for $L < 10$ *km*	(See above)	(See above)	No
[22]	R and SS	*D/AD*	Gamma	None	for $L < 10$ *km*	(See above)	(See above)	Yes
[33]	R	*D/MD*	Gamma	None	location independent	20 global events from Ref. [5]	4.7–8.0	No
[33]	R	*D/AD*	Gamma	None	(See above)	(See above)	4.7–8.0	No
[33]	R	*D/MD*	Gamma	*l/L*	$0 \leq l/L \leq 0.5$	(See above)	4.7–8.0	No
[33]	R	*D/AD*	Gamma	*l/L*	(See above)	(See above)	4.7–8.0	No

* This number includes both L>10 km and L<10 km cases.

TABLE 3. PRINCIPAL FAULT DISPLACEMENT MODELS THAT PREDICT (NON-NORMALIZED) DISPLACEMENT AMPLITUDE.

Reference	Style of Faulting	Predicted Variable	Statistical Distribution	Predictor Variables	Notes	Database	Applicable Magnitude Range	Used in This Study
[14]	SS	$ln(D)$	Normal	M_W, l/L	bilinear profile D in cm $0 \leq l/L \leq 0.5$	13 global events from [37], 9 global events from [14]	6.0–8.0	Yes
[14]	SS	$ln(D)$	Normal	M_W, l/L	quadratic profile D in cm $0 \leq l/L \leq 0.5$	(See above)	6.0–8.0	No
[14]	SS	$ln(D)$	Normal	M_W, l/L	elliptical profile D in cm	(See above)	6.0–8.0	No
[28]	All	$\frac{D^{\lambda}-1}{\lambda}$	Normal	M_W, l/L	D in m $0 \leq l/L \leq 1$ See footnote (1)	73 global events from [5]	6.0–8.0 (SS, N) 5.0–8.0 (R)	Yes
[27]	SS	$ln(D)$	Negative Exponentially Modified Gaussian	M_W, l/L	D in m $0 \leq l/L \leq 0.5$ See footnote (2)	29 global events from [5]	6.0–8.3	Yes
[26]	All	$D^{0.3}$	Normal	M_W, SRL, l/L	D in m $0 \leq l/L \leq 0.5$	74 global events from [5]	5.0–8.5	Yes

[1] Ref. [28] uses a Box-Cox transformation on the predicted variable.

[2] In Ref. [27], the preferred model uses the negative exponentially modified gaussian distribution; alternative models using normal, skew-normal, and skew-t distributions on $ln(D)$ are documented.

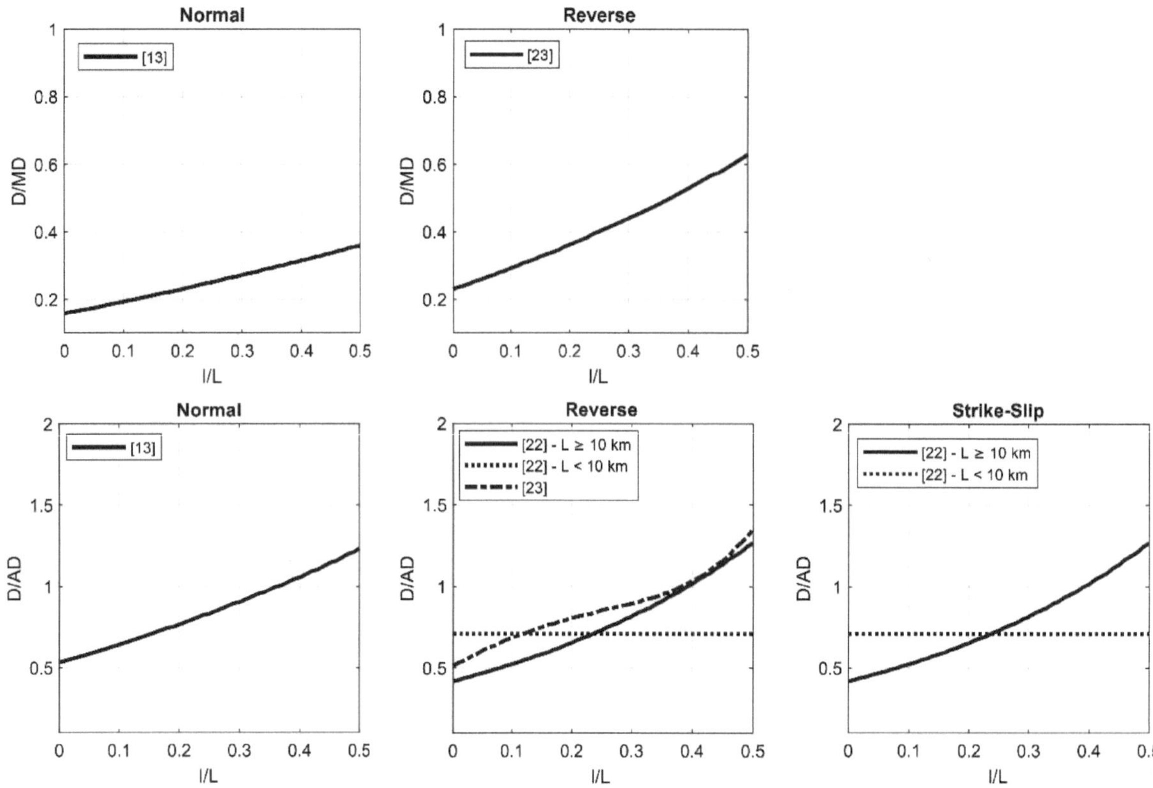

FIG. 5. Median (50th percentile) principal displacement profiles for normalized displacement models (D/MD, top row; and D/AD, bottom row) in Table 2. The results of model proposed by Ref. [22] are the same for reverse and strike-slip styles of faulting.

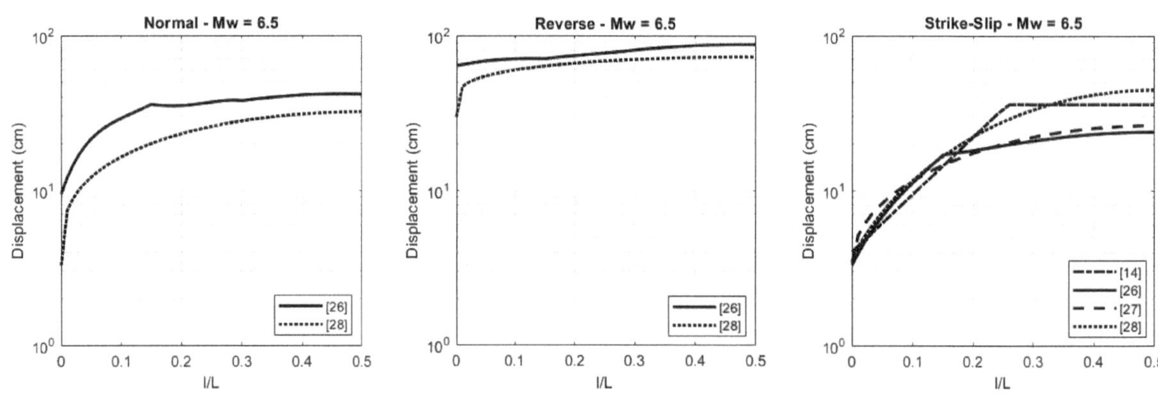

FIG. 6. Median (50th percentile) principal displacement profiles for displacement models in Table 3 using M_W=6.5. For Ref. [28], the mean model coefficients with symmetrical weighting are used (see Ref. [28] for more information). The simplified model (i.e. full rupture length) for aggregate displacement is used for model from Ref. [26].

TABLE 4. DISPLACEMENT METRICS AND APPLICABILITY CONDITIONS FOR FAULT DISPLACEMENT MODELS

Reference	Style of Faulting[1]	Displacement Component	Classification[2]	Used in This Study
[13]	N	Vertical	Single Principal	Yes
[13]	N	Vertical	Single Distributed	Yes
[14]	SS	Lateral	Single Principal	Yes
[14]	SS	Lateral	Single Distributed	Yes
[22]	SS and R	Net	Single Principal	Yes
[23]	R	Vertical	Single Principal	Yes
[22, 24, 25]	SS and R	Net	Single Distributed	Yes
[26]	Any	Net	Aggregate; Sum of Principal	Yes
[27]	SS	Net	Sum of Principal	Yes
[28]	Any	Net	Aggregate	Yes
[33]	R	Vertical	Single Principal	Yes
[38]	R and N	Vertical	Single Distributed	Yes
[39]	R	Vertical	Single Distributed	No

[1] N: Normal, SS: Strike-slip, R: Reverse.
[2] Several newer models provide predictions for the combined discrete displacement on (sub)parallel ruptures, such as sum across principal ruptures or aggregate, which is sum across principal and distributed ruptures (see Section 2.3.2).

3.2.2. Conditional probability of surface rupture on the principal fault

There are a limited number of models available for the conditional probability of non-zero displacement, represented by the $P[D \neq 0|z, r, sr \neq 0]$ term in Eq. (2) [14, 22, 26]. Refs [14, 29]

provide models for the conditional probability that a site of area z^2 is in a gap where no surface ruptures (principal or distributed) occur. Refs [14, 29] also provide models for the aleatory variability in the mapped rupture location ($f_R(r)$) for strike-slip faults. Ref. [26] provide separate models for the conditional probability that no principal ruptures occur at a location l/L and the conditional probability that l/L is in a gap where no surface ruptures (principal or distributed) occur. However, the conditional probability of surface rupture models in Ref. [26] were developed for use with their fault displacement model and are not appropriate for use with other fault displacement models, unless the modelling frameworks are consistent.

Reference [22] provides a model for the conditional probability of non-zero principal displacement for strike-slip and reverse faults based on data from earthquakes in Japan (the P_{2p} term). The P_{2p} term considers the fact that the surface rupture length (indicated by Ls in Fig. 7) is statistically shorter than the source fault length (indicated by Lm in Fig. 7) for small-to-moderate magnitude events. According to Ref. [22], the P_{2p} term can be estimated by using the positional relationship between the surface rupture and the site, and the ratio Ls/Lm. For example, Ls/Lm is approximately 0.31 for a M_W =6.6 earthquake when Line C given in left panel of Fig. 7 is adopted. P_{2p} is calculated as 3/8 (3 out of 8 cases) when the positional relationship between the surface rupture and the site is considered. The final P_{2p} value is obtained as the mean of four estimates given in Lines A–D.

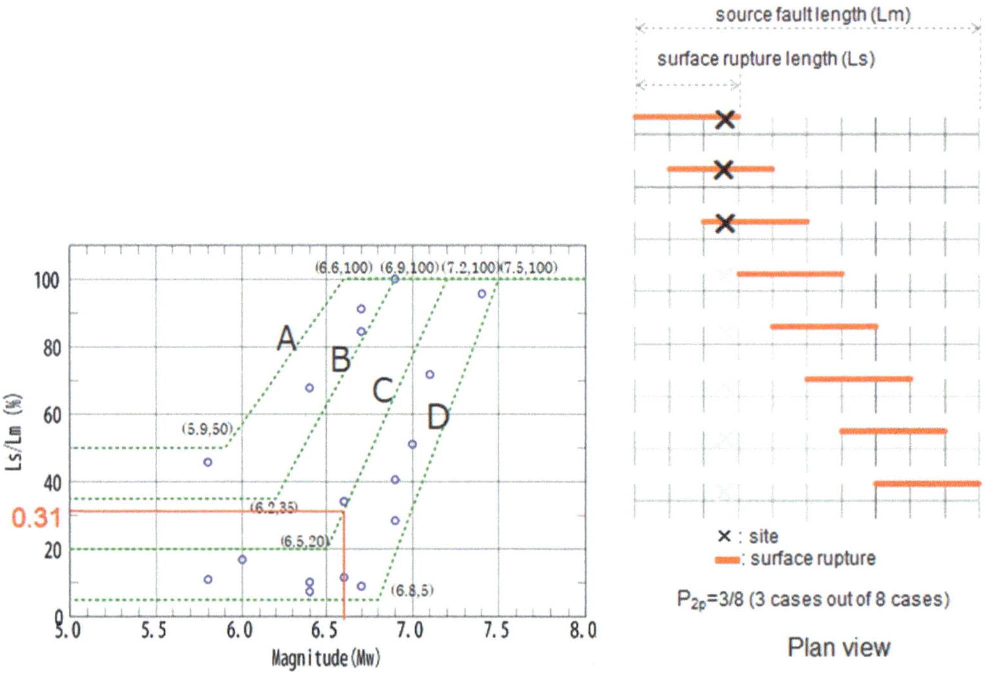

FIG. 7. Left: Alternative regression lines that fit to the empirical data for the relation of ratio of surface rupture length to the source fault length and magnitude for the model proposed by Ref. [22]. Right: example discretization of surface rupture length in hazard calculation.

3.3. DISTRIBUTED FAULT DISPLACEMENT

PFDHA for distributed faulting is calculated using Eq. (3). Available models for the conditional probability of exceedance (i.e. fault displacement models) and conditional probability of surface rupture for distributed faulting are discussed in Sections 3.3.1 and 3.3.2, respectively.

At the time of this publication, there were very few alternative models available for distributed faulting.

Most models for distributed fault displacement hazards are conditioned on the distance r to the principal fault; therefore, the aleatory variability in the mapped principal rupture ($f_R(r)$) is also included in Eq. (3) (Fig. 3(B)). Epistemic uncertainty related to the locational accuracy of the mapped principal fault is sometimes treated as an aleatory variability and included in the $f_R(r)$ term. If no appropriate empirical model is available, this term may be simulated by other methods that evaluate aleatory variability and/or epistemic uncertainty in the modelled location of the principal fault source. In some cases, the distance from the site of interest to the principal fault is much greater than the variability in principal fault rupture location, so the aleatory variability in the mapped rupture might not be significant to the hazard. However, if the site is relatively close to the principal fault, then aleatory variability in the mapped location may be an important contributor to the hazard.

3.3.1. Conditional probability of exceedance on the distributed fault

The conditional probability of exceedance for distributed faulting is treated in a similar way to principal faulting. The conditional probability of exceedance is the $P[d > d_0|r, m, d \neq 0]$ term in Eq. (3). In this term, the probability that the distributed displacement will exceed d_0 at a distance r from the principal surface rupture for a given magnitude m is calculated using distributed fault displacement models. The models are typically empirical and provide a probability density function for the distributed displacement amplitude. Similar to principal fault displacement models, a variety of statistical distributions are used to model the variability (e.g. lognormal, gamma, beta). For simplified applications, formulations for high percentiles as a function of distance r have been provided by some model developers (e.g. 95th percentile in Ref. [13]; 90th percentile in Refs [22, 24, 25]).

Tables 5 and 6 contain non-exhaustive lists of distributed fault displacement models. Models that predict a normalized displacement ratio are listed in Table 5, whereas models that directly predict displacement amplitude are listed in Table 6. As discussed in Section 3.1.1, models that predict a normalized displacement ratio need a separate scaling relation to estimate *MD* or *AD*, and the statistical distribution for the *MD* or *AD* prediction is convolved with the distribution for *d*/*MD* or *d*/*AD* to capture the full variability in the PFDHA calculation. The models are typically developed from empirical data for a specific style of faulting; therefore, the effects of fault kinematics are captured by the data. Figure 8 shows how the median (50th percentile) distributed displacement amplitude attenuates with distance from the principal rupture for the distributed fault displacement models used in this study and listed in Table 5. In all models, the effect of distance is strong but varies significantly with style of faulting. For dip-slip faults, most model developers provide different attenuation relations for the hanging wall and footwall.

TABLE 5. DISTRIBUTED FAULT DISPLACEMENT MODELS THAT PREDICT NORMALIZED DISPLACEMENT.

Reference	Style of Faulting	Predicted Variable	Statistical Distribution	Predictor Variables	Notes	Database	Applicable Ranges	Used in This Study
[13]	N	*d/MD*	Gamma	*r*	Different models for HW and FW	5 global events from Ref.[40]	$5.5 \leq M_W \leq 7.4$ $0 \leq r \leq 15\ km$	Yes
[14]	SS	*ln(d/AD)*	Normal	*r*		8 global events from Ref. [14]	$6.5 \leq M_W \leq 7.5$ $0 \leq r \leq 2.5\ km$	No
[22]	R and SS	*d/MD*	Gamma	*r*		17 Japanese events	$5.8 \leq M_W \leq 7.4$ $0 \leq r \leq 25\ km$	No
[22]	R and SS	*d/AD*	Gamma	*r*		17 Japanese events	$5.8 \leq M_W \leq 7.4$ $0 \leq r \leq 25\ km$	No
[24]	R and SS	*d/AD*	Gamma	*r*		17 Japanese events	$5.8 \leq M_W \leq 7.4$ $0 \leq r \leq 25\ km$	No
[25]	R and SS	*d/AD*	Gamma	*r*		17 Japanese events	$5.8 \leq M_W \leq 7.4$ $0 \leq r \leq 25\ km$	Yes
[33]	R	*d/MD*	Gamma	*r*	Different models for HW and FW. See footnote (1)	20 global events from Ref. [5]	$4.7 \leq M_W \leq 8.0$ $0 \leq r \leq 15\ km$	No

[1] In Ref. [33], *d/MD* represents non-exceedance level of 95% for $r \leq 5.5$ km (hanging wall) and $r \leq 6.5$ km (footwall) and 50% non-exceedance at greater distances; an aleatory variability model is not provided.

TABLE 6. DISTRIBUTED FAULT DISPLACEMENT MODELS THAT PREDICT (NON-NORMALIZED) DISPLACEMENT AMPLITUDE.

Reference	Style of Faulting	Predicted Variable	Statistical Distribution	Predictor Variables	Notes	Database	Applicable Ranges	Used in This Study
[14]	SS	*ln(d)*	Normal	M_W, r	d in cm	8 global events from Ref. [14]	$6.5 \leq M_W \leq 7.5$ $0 \leq r \leq 2.5$ km	Yes
[39]	R	*ln(d)*	Normal	M_W, r, D	Different models for HW and FW	15 global events from Ref. [39]	$4.9 \leq M_W \leq 7.9$ $0 \leq r \leq 1.5$ km	No
[26]	All	d	Parametric model	M_W, d_{tot}, N_{rup}	See footnote (1)	74 global events from Ref. [5]	5.0– 8.5	Yes
[38]	N and R	*ln(d)*	Normal	M_W, r, D	Different models for HW and FW and for type of distributed rupture	34 global events from Ref. [4]	$4.9 \leq M_W \leq 7.9$ $0 \leq r \leq 4$ km (N) $0 \leq r \leq 1.5$ km (R)	Yes

[1] In Ref. [26], a parametric model based on the total budget of distributed displacement (i.e. difference between the aggregate and principal displacement) and number of sub-parallel distributed ruptures.

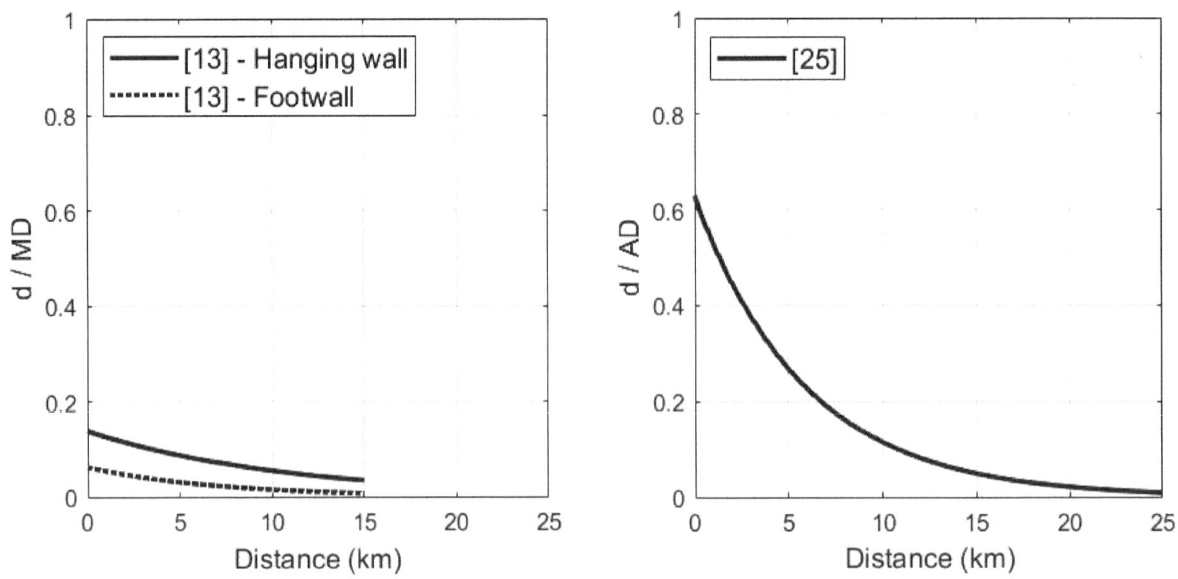

FIG. 8. 50th percentile of distance attenuation of d/MD (left) and of d/AD (right), only for the models used in the IAEA PFDHA exercise.

3.3.2. Conditional probability of surface rupture on the distributed fault

The conditional probability that surface rupture will occur on a distributed fault is handled in the $P[d \neq 0|z, r, sr \neq 0]$ term of Eq. (3). The term is conditioned on surface-fault rupture occurring during the earthquake ($sr \neq 0$), and it provides the probability that a distributed rupture occurs within a site of a specified area z^2 at a distance r given that the principal rupture has occurred.

The currently available conditional probability of rupture models for distributed faulting are constrained by empirical data. A common methodology to develop the distributed rupture probability models is to overlay a grid with cell area z^2 on a rupture map and calculate the frequency of cells containing one or more distributed ruptures located at a distance r from the principal rupture [13, 14, 22, 24, 41]. More recently, Refs [38, 39] used fault-parallel slices instead of square cells to evaluate the frequency of distributed ruptures.

Table 7 is a non-exhaustive list of models for the conditional probability of distributed surface rupture. Most models only consider the effects of distance r from the principal rupture. While many models have earthquake magnitude as a predictor variable, some model developers found no strong dependence on earthquake size [14, 24, 41]. Most models are developed from empirical data for a specific style of faulting; therefore, the effects of fault kinematics are captured by the data selection. For dip-slip faults, some model developers provide different attenuation relations for the hanging wall and footwall. References [14, 24] provide a set of models for a wide range of cell areas (z^2).

Figure 9 displays the conditional probability of distributed fault rupture as a function of r for the subset of models evaluated in this exercise, sorted by style of faulting. All models show a decrease in probability with increasing distance r from the principal fault. The models developed by Refs [13, 38] include a correlation with magnitude, with higher probabilities associated with larger magnitudes.

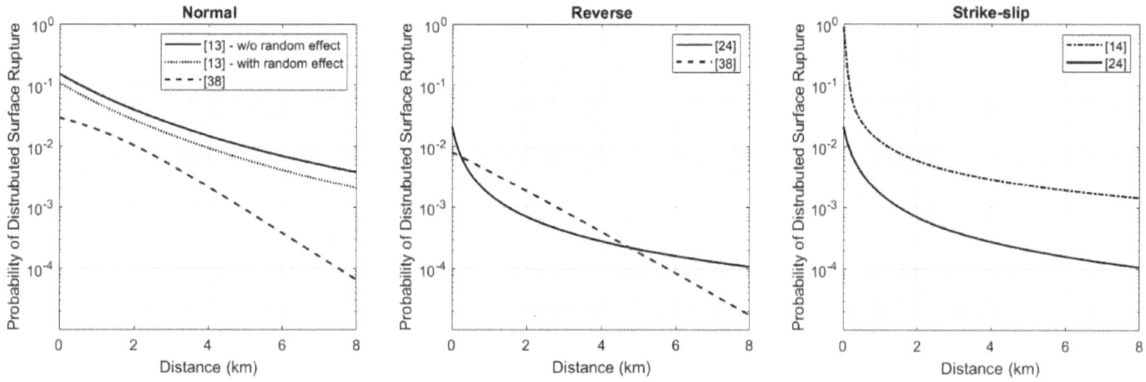

FIG. 9. Conditional probability of slip for distributed faulting for different kinematics. Profiles from Refs [14, 24, 38] are for a 100m x 100m grid size. The curves of Refs [13, 38] are computed for M_w = 6.7. All models are for the hanging wall only. Model from Ref. [38] is for distributed ruptures ranked as 2 (see Ref. [38] for more details).

3.4. FIRST-ORDER INTERPRETATION OF HAZARD CURVES

Many of the key differences in the results provided in the next sections are related to the use of different conditional probability of surface rupture models for the earthquake (Section 3.1) or for principal or distributed faulting models (Sections 3.2.2 and 3.3.2, respectively). Example hazard curves are provided in this section to assist readers in understanding the results from different hazard analysts by demonstrating how the conditional probability of surface rupture models affect the hazard. The example uses a principal fault displacement model to define the conditional probability of exceedance on a principal fault (Section 3.2.1), and an event conditional probability of surface rupture model (Section 3.1). In these examples, the fault source characterization is simplified by assuming earthquake recurrence of a single magnitude at a mean rate.

The left plot in Fig. 10 shows example hazard curves for cases of M_W=5.5 (in blue) and M_W=6.5 (in orange) earthquakes, occurring beneath the site at a mean rate of 0.002 a^{-1} (equivalent to a mean recurrence interval of 500 years). No model for the conditional probability of surface rupture has been applied, so a conditional probability of 1.0 is implied (all earthquakes will produce surface-fault rupture). The hazard curves are calculated using the bilinear model of Ref. [14] (Table 3) with a fixed along-rupture distance ratio l/L = 0.4. The intersection of the hazard curves with the y-axis marks the annual frequency of exceedance for a small (1 cm) displacement. Consistent with the mean rate of earthquakes in the source model, the hazard curves for both cases intersect the y-axis at approximately 2×10^{-3} a^{-1}.

TABLE 7. PROBABILITY OF SURFACE RUPTURE MODELS FOR DISTRIBUTED FAULTING.

Reference	Style of Faulting	Function Type	Predictor Variables	Cell Sizes (m)	Notes	Applicable Ranges	Used in This Study
[13]	N	Logistic	M_W, r, *flag*	500 × 500	HW/FW flag; Model without random effects	$5.5 \leq M_W \leq 7.4$ $0 \leq r \leq 15$ km	Yes
[13]	N	Logistic	M_W, r, *flag*	500 × 500	HW/FW flag; Model with random effects	(See above)	Yes
[14]	SS	Power	r	25 × 25, 50 × 50 100 × 100, 150 × 150 200 × 200	Footnote (1)	$6.5 \leq M_W \leq 7.5$ $0 \leq r \leq 2.5$ km	Yes
[22]	R and SS	Logistic	M_W, r	500 × 500		$5.8 \leq M_W \leq 7.4$ $0 \leq r \leq 25$ km	No
[24]	R and SS	Logistic	r	50 × 50, 100 × 100 250 × 250, 500 × 500		(See above)	Yes
[39]	R	Logistic	M_W, r	10 × ∞	Cells are strike-parallel	$4.9 \leq M_W \leq 7.9$ $0 \leq r \leq .5$ km	No
[41]	N	Logistic	r	500 × 500	Different models for HW and FW; regular model	$6.0 \leq M_W \leq 7.5$ $0 \leq r \leq 15.5$ km (HW) $0 \leq r \leq 12.5$ km (FW)	No
[41]	N	Logistic	r	500 × 500	Different models for HW and FW; conservative model	$6.0 \leq M_W \leq 7.5$ $0 \leq r \leq 15.5$ km (HW) $0 \leq r \leq 12.5$ km (FW)	No
[33]	R	Constant	None	100 × ∞ 500 × ∞	Different models for HW and FW; Cells are strike-normal	$6.6 \leq M_W \leq 7.9$ $0 \leq r \leq 3$ km	No
[33]	R	Fit to Cumulative Distribution	r	Not Applicable	Different models for HW and FW; Different models based on magnitude	(See above)	No
[33]	R	Constant	None	100 × ∞ 500 × ∞	Different models for HW and FW; Cells are strike-normal	$6.0 \leq M_W \leq 8.0$ $0 \leq r \leq 15$ km	No
[33]	N	Constant	None	100 × ∞ 500 × ∞	(See above)	$6.2 \leq M_W \leq 7.8$ $0 \leq r \leq 15$ km	No
[33]	R	Fit to Cumulative Distribution	r	Not Applicable	Different models for HW and FW; Different models based on magnitude	$6.0 \leq M_W \leq 8.0$ $0 \leq r \leq 15$ km	No

Reference	Style of Faulting	Function Type	Predictor Variables	Cell Sizes (m)	Notes	Applicable Ranges	Used in This Study
[38]	N and R	Logistic	M_W, r	$10 \times \infty$		$4.9 \leq M_W \leq 7.9$ $0 \leq r \leq 4$ km (N) $0 \leq r \leq 1.5$ km (R)	Yes
[26]	All	Composite Half-Normal and Uniform	M_W, r	100 m		$5.0 \leq M_W \leq 8.5$ $0 \leq r \leq 20$ km	Yes

[1] In Ref. [14], near-field adjustments are applied as described in the publication to correct for power function extrapolation errors.

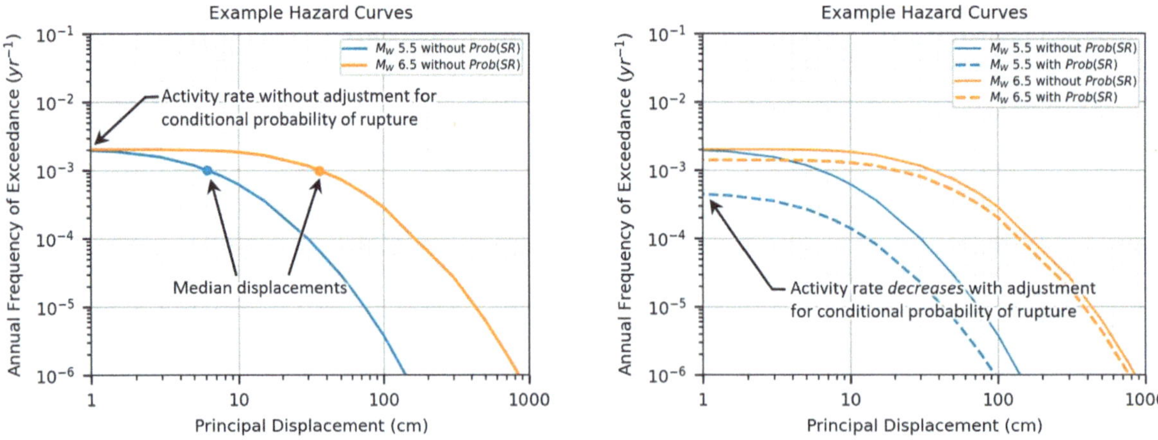

FIG. 10. Examples provided with discussion (see text) for first-order interpretation of fault displacement hazard curves.

As expected, the hazard (i.e. annual frequency of exceedance) is higher for the larger magnitude case at all displacements. The median displacement (marked by a dot on the hazard curve) for the M_W=6.5 case is greater than the median for the M_W=5.5 case. The slope of the hazard curve is controlled by the aleatory variability and the number of standard deviations above or below the median. Both curves in the left plot in Fig. 10 have the same aleatory variability; therefore, the slopes at a given statistical level above or below the median are the same (which might be obscured by the log scale).

The right plot in Fig. 10 shows hazard curves with and without the conditional probability of surface rupture. The solid lines are hazard curves without the conditional probability (i.e. identical to the left plot), and the dashed lines include the conditional probability of surface rupture model of Ref. [20] based on a global data set (Table 1 and Fig. 4). Application of the model yields 23% and 70% probabilities of surface rupture for the M_W=5.5 and M_W=6.5 cases, respectively. Including the conditional probability of surface rupture linearly scales the hazard curve down because it impacts the mean rate (or frequency) of events that rupture the ground surface. The impact is small for the higher magnitude M_W=6.5 case but is significant for the lower magnitude M_W=5.5 case.

The example in Fig. 10 demonstrates the first-order results from PFDHA. Fault displacement models that predict higher median displacements for a given magnitude will 'stretch' the hazard curve (on log-log plots) farther to the right along the displacement axis relative to models that that predict lower median displacements. Fault displacement models that have lower aleatory variability will have steeper hazard curves at comparable percentiles above the median displacement compared to models with higher aleatory variability. The conditional probability of surface rupture is a function of magnitude in the currently available empirical models, and most models produce much lower probabilities of surface rupture for lower magnitude earthquakes compared to higher magnitudes. Because the probability of surface rupture scales down the hazard curve along the y-axis, hazard curves based on smaller magnitude earthquakes will be more affected by this term than hazard curves based on larger magnitude earthquakes. Although the example in Fig. 10 is for principal faulting, the same concepts apply to distributed faulting: larger median displacements move the hazard curve to the right, less aleatory variability produces a steeper curve, and lower probabilities of surface rupture move the curve down.

4. DESIGN OF THE PROBABILISITIC FAULT DISPLACEMENT HAZARD ANALYSIS EXCERCISE

The IAEA PFDHA exercise started officially in November 2020. A flowchart showing the project implementation is provided in Fig. 11. The first step (Step 0) of the project included the selection of the external experts to form the Advisory Board and establishing the primary project objectives and milestones. The Advisory Board includes representatives from External Events Safety Section of IAEA and external experts representing the utilities, technical support organizations, and sector professionals. The Advisory Board defined the important milestones of the project as follows:

— Step 1: Utilization of published PFDHA models for a single case to compare the primary and distributed fault displacement hazard curves;

— Step 2: Enlarging the scope of the project by adding more cases to represent different fault mechanisms and inviting additional modelling teams to involve the new PFDHA models to be published in 2023 and 2024;

— Step 3: Documentation of the results including hazard code verification, comparison of the hazard curves and dissemination of the results.

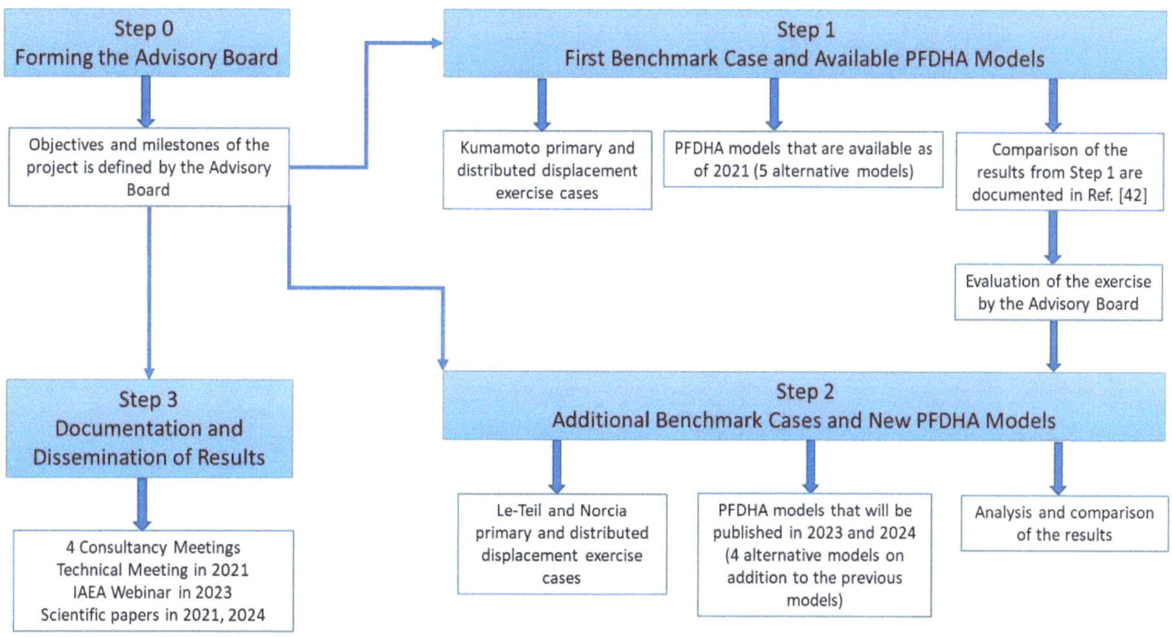

FIG. 11. Workflow of the IAEA PFDHA exercise.

4.1. STEP 1: FIRST BENCHMARKING CASE AND AVAILABLE FAULT DISPLACEMENT MODELS

The primary objective of Step 1 was to establish the baseline for the project using the available PFDHA models to identify the differences in the hazard results and the challenges for designing the benchmark cases. For this purpose, the Advisory Board recommended:

— To start with a single case study, the 2016 Kumamoto earthquake;

— Engage the authors of the published PFDHA models and request that they provide the hazard curves for principal and distributed displacement for this case study;

— Analyse differences in the hazard curves and evaluate the needs for proper parametrization of benchmark cases for Step 2;

— Establish the first set of baseline hazard curves for following verification exercises.

The scenario used in this step was based on the M_W=7.0 strike-slip Kumamoto earthquake that occurred in 2016. The details of the case study are provided in Section 5. In this step, a relatively simple scenario that includes simplified seismic source parameterization was preferred. To ensure the consistency among hazard analyst teams, all seismic source model parameters were provided by the Advisory Board. In particular, the site coordinates and dimensions, and the full source characterization logic tree were given to the modelling teams in a spreadsheet and in Esri shapefile format.

Five hazard analyst teams were invited to participate to the exercise in this step. These participants shared their analysis results with the Advisory Board and discussed the results in an interactive environment facilitated during the consultancy meetings. Step 1 of the project was finalized in April 2021 and the preliminary results were published in Ref. [42].

4.2. STEP 2: ADDITIONAL BENCHMARKING CASES AND NEW FAULT DISPLACEMENT MODELS

The analysis performed in Step 1 showed significant scatter among the results from different hazard analyst teams. The Advisory Board recommended the inclusion of other hazard analyst teams with significant experience in implementing PFDHA. Inviting new PFDHA modelling teams to participate in this project allows:

— Multiple models for each style of faulting, leading to more robust model comparisons;

— The opportunity to utilize recent models published for specific components of PFDHA;

— Verifications to ensure published PFDHA models are implemented correctly.

In addition to the modellers who participated in the first step, three more lead modellers were involved in Step 2. Furthermore, two modellers were able to update the models that they used in Step 1. Another group of experts have joined the benchmark exercise in Step 2, in order to carry out the verification exercise that ensures the published PFDHA models were implemented correctly in the hazard codes.

A technical meeting on Benchmarking of Current Practices in Probabilistic Fault Displacement Hazard Assessment for Nuclear Installations was held in November 2021. Discussions in that meeting were evaluated by the Advisory Board. The Advisory Board recommended that more cases need to be added to the benchmark exercise with the aim of using only the suitable models for each style of faulting. Therefore, in addition to the Kumamoto case for strike-slip faulting, principal, and distributed displacement cases for normal faulting (2016 Norcia earthquake) and reverse faulting (2019 Le Teil earthquake) were designed. Details of these new cases are provided in Sections 6 and 7, respectively. Similar to the Kumamoto case, the Advisory Team provided the seismic source model parameters for the principal and distributed fault displacement exercises to model the epistemic uncertainty in different geometric and kinematic parameters, and to investigate the impact of different input parameters on the resultant final hazard.

Three different exercises for each style of faulting were designed and parameterized for the assessment of distributed fault displacement, considering the fault capability constraints in the site vicinity of a nuclear installation. The site to fault distances in the distributed fault displacement exercises range from 0.6 to 7.6 km and the causative sources are the ones that have clearly shown evidence of surface rupture. These exercises were designed to explore the uncertainty in some key parameters, such as the moment rate, slip rate, M_W, dip direction, rupture width in the geometry and kinematics of the capable faults. The exercises were also tailored to understand the impact in the final hazard curves when, for example, alternative fault source or a different site-fault distance are considered.

Step 2 of the project included three principal and nine distributed fault displacement exercises. The hazard analyst teams comprised of more than twenty experts (researchers and practitioners in the field of PFDHA) and provided hazard curves for nine different PFDHA models. Due to the type of each fault displacement model, not all were able to perform both principal and distributed fault displacement exercises. For this reason, depending on the model, the Advisory Board was able to collect the principal and all distributed exercises for some PFDHA models, or only the principal fault displacement exercise or only all distributed fault displacement cases from other PFDHA models. Table 8 provides a summary of the collected results (i.e. hazard curves) from each model. The comparisons of the hazard curves for these analyses are provided in Sections 5–7.

TABLE 8. MATRIX OF THE MODELLING TEAMS AND CASE STUDIES (THE CELLS WITH X SHOWS THE AVAILABLE ANALYSIS RESULTS).

Model Abbreviation	Principal	1. Distributed	2. Distributed	3. Distributed	Reference
Kumamoto Case Study					
C24	X	-	-	-	[27]
L23	X	-	-	-	[26]
K24	X	-	-	-	[28]
P11	X	X	X	X	[14]
T13	X	X	X	X	[22, 24, 25]
Le Teil Case Study					
L23	X	-	-	-	[26]
K24	X	-	-	-	[28]
M11	X	-	-	-	[23]
T13	X	X	X	X	[22, 24, 25]
V24	-	X	X	X	[38]
Norcia Case Study					
L23	X	-	-	-	[26]
K24	X	-	-	-	[28]
V24	-	X	X	X	[38]
Y03	X	X	X	X	[13]

4.3. STEP 3: PROJECT DOCUMENTATION AND DISSEMINATION OF THE RESULTS

As the overall objective of the project is to support the Member States performing PFDHA for nuclear installations sites, the most relevant outcome of the project is this publication. In addition, two scientific papers were prepared for the technical audience [42, 43] based on the results of Step 1 and Step 2, respectively. The project aimed to create an interactive environment among the hazard analyst teams; therefore, 4 consultancy meetings were organized between November 2020 and May 2023. Results of Step 1 were disseminated to and discussed among a larger group during the Technical Meeting held in November 2021. On 30 May 2023, a webinar entitled 'Recent Advances in Probabilistic Fault Displacement Hazard Assessment for Nuclear Installations in Light of Geological Reconnaissance Findings' was organized to disseminate the project results [44].

5. THE STRIKE-SLIP FAULTING CASE STUDY: KUMAMOTO

In April 2016, the Kumamoto Prefecture (Kyushu Region, Japan) was struck by an earthquake sequence that began on the 14th of April with a M_W=6.2 foreshock. Two days later, on the 16th of April, a M_W=7.0 mainshock hit Kumamoto city, yielding a ~30 km long surface rupture starting from the eastern part of the Futagawa fault zone[1] and reaching up to the northernmost part of the Hinagu fault zone [45].

The Futagawa and the southern part of the Hinagu fault zones are parts of the ENE-WSW striking Oita-Kumamoto and the Usuki-Yatsushiro tectonic lines, respectively. These fault zones are located in the middle of Kyushu Island and is the area affected by the 2016 Kumamoto earthquake sequence. The Futagawa fault zone consists of three segments: Futagawa, Uto, and Uto-Hanto-North, as shown in Fig. 12. The Hinagu fault zone (Fig. 12), whose northern part links the Usuki-Yatsushiro tectonic line with the Oita-Kumamoto tectonic line, also consists of three segments according to its rupture history and geometry: the Takano-Shirahata, Hinagu, and Yatsushiro Sea segments [46].

After the 2016 earthquake sequence, several researchers published seismic source models based on the observed crustal deformation and/or inversion analyses of recorded strong ground motions [47–50]. Most of these models agreed that the M_W=6.2 foreshock had occurred on a vertical fault plane along the Takano-Shirahata segment, and it was characterized by right-lateral strike-slip motion. The M_W=7.0 mainshock had occurred on a 60–70° NW-dipping fault plane and partially ruptured the Uto and Futagawa segments. The mechanism of the mainshock was a right-lateral strike-slip with a normal component. The rupture plane on the northern part of the 70–80° WNW-dipping Takano-Shirahata segment was characterized as an almost pure right-lateral strike slip.

In addition to the fault displacement on the principal fault associated with the M_W=7.0 Kumamoto earthquake, several small, distributed ruptures were documented, in some cases even 10 km away from the principal fault trace on Futagawa fault zone [51, 52]. Reference [52] describes three subparallel NW-trending fault strands with ~10-cm-scale co-seismic offset or deformation along the Suizenji fault and its extension.

5.1. PRINCIPAL AND DISTRIBUTED FAULT DISPLACEMENT EXERCISES

One principal and three distributed displacement exercise cases were developed, and a full set of parameters were provided to the fault displacement hazard analysts to explore the effect of the magnitude and source-to-site distance on the results (Table 9). The site dimensions (z) were defined as 100 × 100 m for all (principal and distributed displacement) cases. The seismic source parameters summarized in Table 10 were selected based on the source parameters of the Kumamoto earthquake sequence proposed by Headquarters for Earthquakes Research Promotion of Japan [53] and reviewed by the Advisory Board. Both for the principal and distributed fault displacement exercises, only the Futagawa fault zone and the Suizenji fault were included in seismic source characterization. The Hinagu fault zone was excluded, due to its long distance from the selected sites. The fault length of each segment was determined by the Geological Survey of Japan, considering geomorphological and geological features and the uncertainty in the fault length is not considered for this exercise.

[1] In this publication, the term 'fault zone' refers to the multiple segments in a fault system.

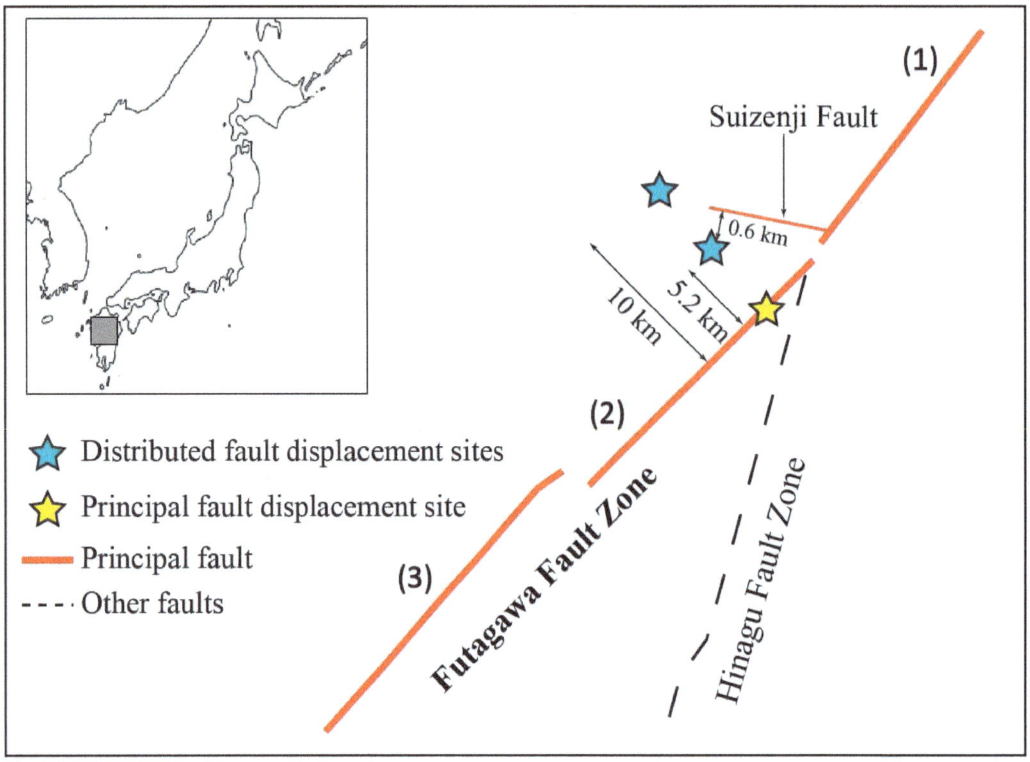

FIG. 12. Sketch of Futagawa fault zone and Suizenji fault, and the site locations for principal and distributed fault displacement exercises for the strike-slip case. The grey square in the insert shows the location of the site area. Segments of Futagawa fault zone are denoted by Futagawa (1), Uto (2), and Uto-Hanto-North (3).

TABLE 9. SUMMARY OF PARAMETERS USED FOR KUMAMOTO PRINCIPAL AND DISTRIBUTED FAULT DISPLACMENT EXERCISES.

Cases	*r* (km)	Site Coordinates	Sources included in PFDHA	*l/L*	Map Accuracy
Principal Option 1	0	130.766–32.747	Uto Segment	0.23	Not applicable
Principal Option 2 (Floating)	0	130.766–32.747	See Table 10	0.1–0.37	Not applicable
1. Distributed	5.2	130.742–32.79	Futagawa Fault Zone	0.1–0.37	Approximately Located
2. Distributed	10	130.72–32.829	Futagawa Fault Zone	0.1–0.37	Approximately Located
3. Distributed	0.6	130.742–32.79	Suizenji Fault	0.39	Inferred

In the principal fault displacement case, the site was assumed to be located on the principal fault, at the location indicated by the yellow star in Fig. 12. The hazard analysts were given two options for principal PFDHA runs. In the first option, a single-segment rupture scenario that includes only the Uto segment is considered. The magnitude of the event was calculated as M_W=6.5, based on the rupture dimensions of the Uto segment, using the magnitude scaling relationship given in Refs [54, 55]. In this option, the along-rupture distance ratio, *l/L* is 0.23. In the second option, the analysts were asked to float the rupture length of the M_W=6.5 earthquake (~20 km) along the Futagawa fault zone, over the full length of the three segments (Segments 1+2+3, Fig. 12). In this case, *l/L* varies according to the location of each rupture along the full length of the fault. For both options, other geometrical parameters such as dip angle and seismogenic thickness (Table 9), were inferred from literature [46, 53].

TABLE 10. SEISMIC SOURCES AND PARAMETERS USED FOR THE STRIKE-SLIP CASE STUDY.

Segments	Maximum Rupture Length (km)	Maximum Rupture Thickness (km)	Average Dip (°)	Magnitude	Mean Rate of Occurrence ($\times 10^{-5}$ a^{-1})
Uto (2)	22	14	60 NW	6.5	18.9
Futagawa (1) + Uto (2)	46	14	68.6 NW	6.9	1.28
Uto (2) + Uto-Hanto-North (3)	54	14	60 NW	7	3.53
Futagawa (1) + Uto (2) + Uto-Hanto-North (3)	78	14	64.6 NW	7.2	1.28
Suizenji	5.4	14	60 SW	5.8	23.30

For defining the occurrence rate, the average recurrence interval was calculated based on the average slip rate and dislocation per event. For this calculation, the scaling relationship between fault length and dislocation given in Ref. [56] was used to compute the dislocation per event. The average slip rate was determined from geomorphological survey results provided by Ref. [53]. Based on these constraints, the mean annual rate of occurrence for the M_W=6.5 earthquake was calculated as 1.89×10^{-4} a^{-1} for the single-segment scenario (Option 1) and 6.42×10^{-4} a^{-1} for the floating option (Option 2). However, the actual (or net) mean annual rate of occurrence for the floating case is quite close to that of the single-segment case, as the mean annual rate of occurrence is normalized by the probability of the rupture passing by the site (a set of 22 km long ruptures may be floated along the strike of the 78 km long fault system) in the hazard runs. It is noteworthy that all parameters were assigned single values, in other words, the epistemic uncertainty in fault length, fault width, dip angle, mean annual rate of occurrence, and the aleatory variability in the magnitude of the ruptures are discarded for simplicity.

Three distributed fault displacement cases were developed with different M_W and source-to-site distance (r) values. For the first distributed fault displacement exercise, a combination of the three segments of the Futagawa fault zone (Segments 1+2+3) was considered. The source-to-site distance r was selected as 5.2 km, and the site is located closer to the northeast end of the Uto segment as shown by the blue star in Fig. 12. For this case, hazard analysts were asked to run the analysis once for single values for each parameter (single path) and once by using the logic tree given in Fig. 13 (multi path). Therefore, the epistemic uncertainty in source model parameters was ignored in the single path case and the uncertainty in the magnitude and rate of occurrence was considered in the multi path case. In addition, variability in the rupture source (single or multi segment ruptures) was included in the logic tree for the multi path case. The magnitude and mean rates of occurrence for the centre branches of the logic tree given in Fig. 13 are the same as in Table 10. The magnitude uncertainty (±0.2) reflects a factor of 2 uncertainty in the seismic moment per event, and the uncertainty in the mean occurrence rate is a factor of 3 with respect to the centre branch.

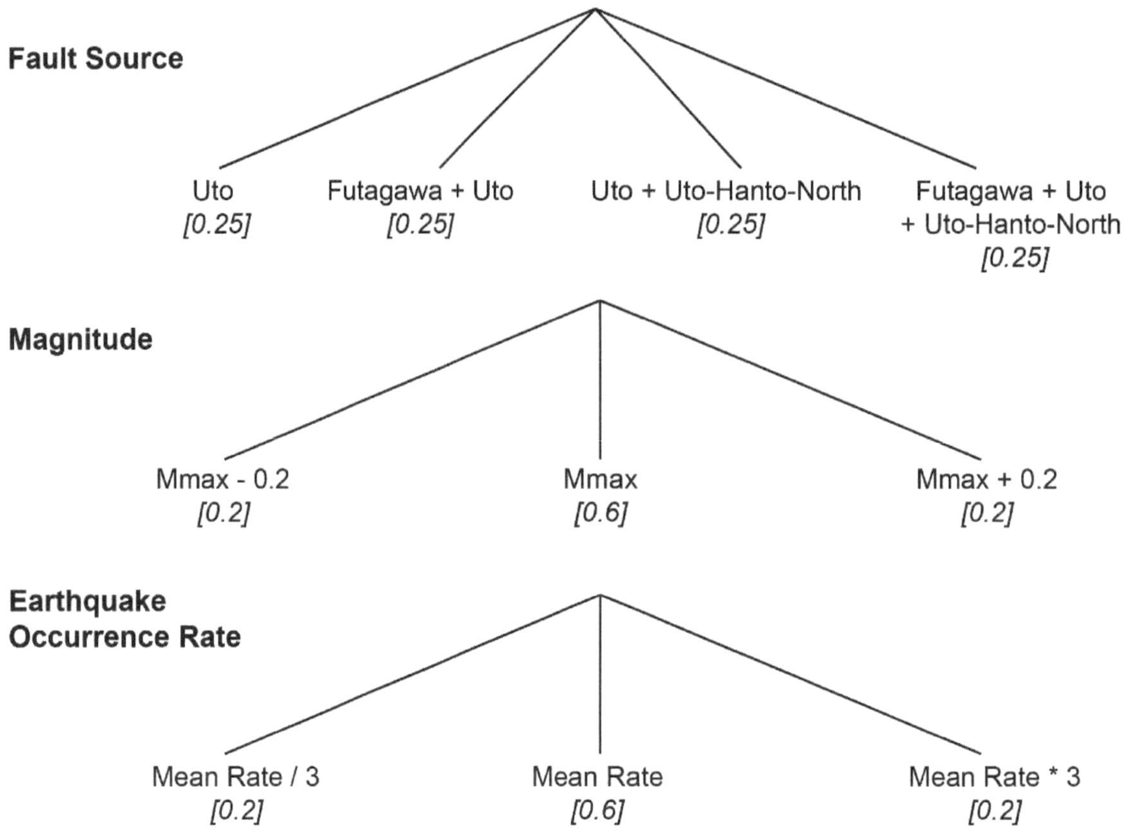

FIG. 13. Logic tree used in the multi path option of the first distributed fault displacement exercise. The weight of each branch is shown in square brackets. The magnitude and mean occurrence rate of the centre branch for each scenario are listed in Table 10.

The second distributed fault displacement exercise was almost the same as the to the first one, with a single path, except that the site-to-source distance, *r*, was increased to 10 km (shown by the blue star in Fig. 12).

The site for the third distributed fault displacement exercise was the same as the first distributed fault displacement exercise (Fig.12). However, the Suizenji fault (Fig. 12) was treated as the principal fault source for this case so the closest distance to the fault is 0.6 km. Suizenji fault was not identified by Ref. [53] before the 2016 Kumamoto earthquake; therefore, the fault length and geomorphological survey results published in Ref. [52] were used to compute the magnitude and mean occurrence rate for this scenario, following the same procedure described above. The mapped accuracy was inferred and the epistemic uncertainty in source parameters was not included in this case (Table 10).

5.2. COMPARISON OF THE HAZARD RESULTS

Five hazard analyst teams provided fault displacement hazard curves for the exercises developed for the Kumamoto case, using the PFDHA models listed in Table 8. Two teams were able to provide the results for all cases. However, due to the limited number of fault displacement models applicable for distributed fault displacement, three teams could only run the PFDHA for the principal fault displacement case. Figures 14 and 15 compare the PFDHA results for the principal and distributed fault displacement exercises, respectively. The hazard curves in these figures are expressed in terms of annual frequency of exceedance (AFOE, in a^{-1}) versus fault displacement (in cm).

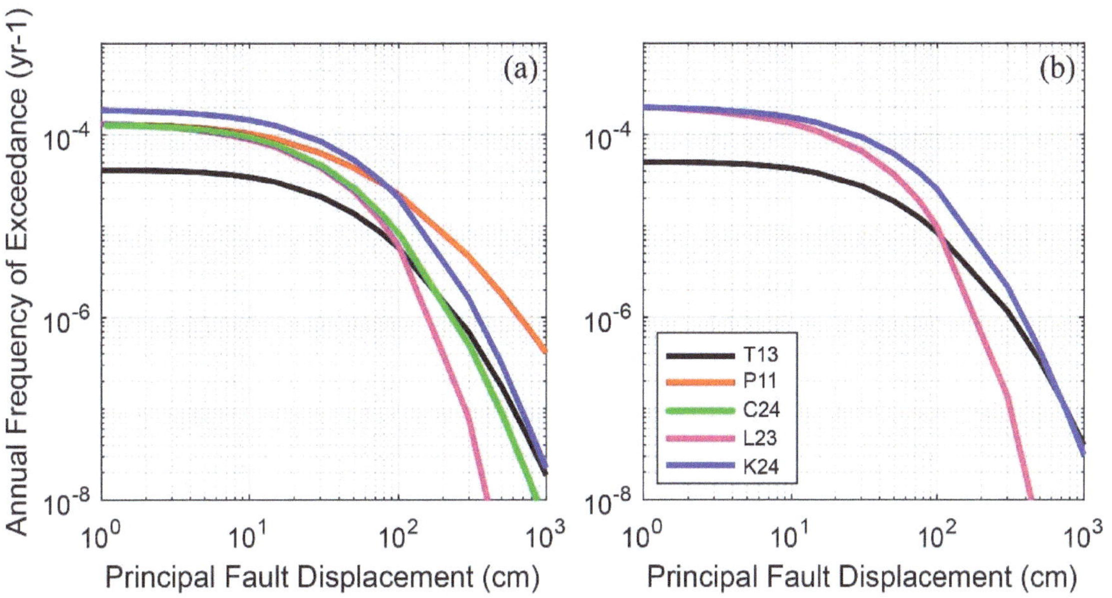

FIG. 14. Principal fault displacement hazard curves in terms of annual frequency of exceedance (AFOE) versus fault displacement for the principal fault displacement exercise. The two frames are for the result (a) without and (b) floating ruptures (Options 1 and 2, respectively). For model abbreviations, please refer to Table 8.

In the principal fault displacement case, the hazard curves are generally in good agreement for the options with and without the floating ruptures, as shown in Fig. 14. The differences in the y-axis of the fault displacement hazard curves are mainly driven by the surface rupture probability models adopted by different analysts (see Sections 3.1 and 3.4). Hazard analysts that implement the K24 model did not incorporate the conditional probability of principal surface rupture (the surface rupture probability was assumed to be 1), resulting in the highest hazard estimates for both options. The hazard estimates of the PFDHA runs that implemented the P11 and C24 models were slightly lower, as the hazard analysts used the surface rupture probability model given in Ref. [20], which yielded a probability of 0.7 for M_W=6.5 earthquakes. In the hazard analysis that utilized the L23 model, the case-specific surface rupture probability was estimated by floating the rupture plane over the dip-slip surface in the hazard runs. The hazard estimates based on the T13 model is the lowest for both options. This is because the probability of the surface rupture for a M_W=6.5 earthquake is approximately 45% in the surface rupture probability model of Ref. [22] (Fig. 4). In addition, the T13 hazard curves were further scaled down, considering the conditional probability of P_{2p}, as explained in Section 3.3.2.

For almost all fault displacement models, estimated principal fault displacements for a 10 000 year return period (AFOE=10^{-4}) are in the range 6–22 cm without the floating ruptures, and 15–26 cm for floating ruptures. Due to the additional P_{2p} factor, the annual probability of the principal fault displacement is less than 10^{-4} for the M_W=6.5 event when the T13 PFDHA model is utilized.

Figure 15 compares the distributed displacement hazard curves (top row is for the first distributed displacement case with Fig. 15(a) showing the single path and Fig. 15(b) showing the multi path results). When compared to the principal fault displacements for the same AFOE given in Fig. 14, the distributed fault displacements are significantly low. Comparing Fig. 15(b) to Fig. 15(a) emphasizes that the mean hazard curves are almost the same, as the centre of the multi path logic tree is the same as the single path case. However, including the epistemic uncertainty in source parameters results in a factor of 10 change in distributed fault displacement estimates within the 90% confidence interval. Among the two distributed PFDHA models that are applicable to strike-slip mechanism, the P11 model represents the upper bound up to 40 cm distributed fault displacement. For distributed displacements higher than 40 cm, estimates of the T13 model are higher. However, the annual rate of exceedance for displacements over 40 cm is quite low (AFOE<10^{-8}). Two important reasons for the differences in the estimates of two hazard analysis teams are: (i) differences in the conditional probability of distributed surface rupture among the models proposed by Refs [14, 22] shown in Fig. 9, and (ii) differences in the standard deviation of the distributed fault displacement models. The source-to-site distance for this exercise (5.2 km) is out of the range of applicability for the P11 model; therefore, the distance attenuation model of distributed surface rupture proposed by Ref. [14] had to be extrapolated for the first and second distributed displacement cases. Differences in the standard deviations of the models affect the slopes of the hazard curves, resulting in the change of the upper bound model for very low AFOEs.

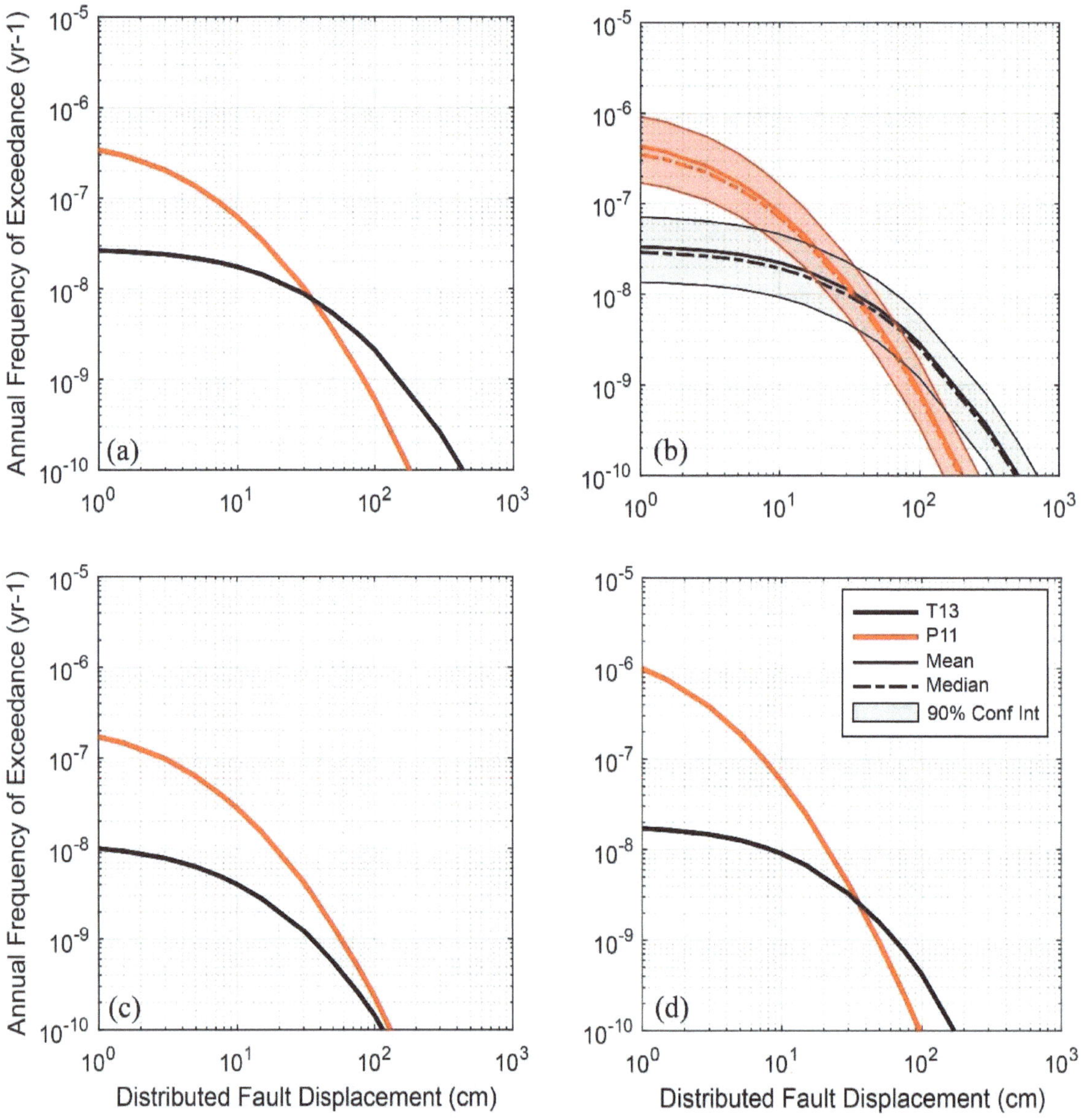

FIG. 15. Distributed displacement hazard curves in terms of annual frequency of exceedance (AFOE,) versus distributed fault displacement. (a) 1st distributed displacement case (r=5.2 km), results of single path, (b) 1st distributed displacement case (r=5.2 km), results of multi path, (c) 2nd distributed displacement case (r=10 km), and (d) 3rd distributed displacement case (r=0.6 km). For model abbreviations, please refer to Table 8.

In the second distributed fault displacement case (with *r*=10 km), the P11 PFDHA model represents the upper bound for all displacement values. The difference in the displacement amplitudes for the same AFOE is again dominated by the conditional probability of distributed surface rupture shown in Fig. 8 and standard deviations associated by the fault displacement models. When Fig. 15(c) is compared to Fig. 15(a), a small difference in the estimations of P11 model is observed, even if the source-to-site distance is significantly increased. It's important to note that both exercise cases have source-to-site distances far beyond the range of applicability for this model; therefore, the underlying reason for this small difference is difficult to comprehend. On the other hand, the distributed fault displacement estimates of T13 model decreases with increasing *r*: the AFOE for 30 cm of distributed fault displacement decreases from 10^{-8} a^{-1} in Fig. 15(a) to 10^{-9} a^{-1} in Fig. 15(c).

In the third distributed fault displacement exercise (Fig. 15(d)) the source-to-site distance is equal to 0.6 km. In this case, the Suizenji fault was treated as the primary source of hazard, which has a lower magnitude compared to the first case single path analysis (M_W=5.8 vs. M_W=6.5) but the same mean annual rate of earthquake occurrence was utilized. The hazard curve for one of the models (P11) is significantly higher when compared to Fig. 15(a), indicating that the effect of source-to-site distance on the PFDHA results is significant. On the other hand, the difference between the hazard curves increased significantly, as the estimates of the T13 model did not change very much in comparison to the first distributed displacement case.

The main conclusions and the lessons learned from the Kumamoto exercise cases are discussed in Section 8.

6. THE REVERSE FAULTING CASE STUDY: LE TEIL

On 11 November 2019, the Le Teil town in southeast-central France was struck by the strongest (M_W=4.9) and most destructive earthquake ever felt in France since the Arette earthquake in 1967 [57, 58]. Despite its small magnitude, the 2019 Le Teil earthquake yielded to a ~5 km long surface rupture on the NE-SW trending La Rouvière fault. The surface rupture occurring in such a small magnitude event is possibly related with its shallow hypocentral depth, which was inferred as 1–3 km by the analysis of the recorded waveforms [59].

The La Rouvière fault is located within the north-eastern termination of the Cévennes fault zone, which acts as a major structural boundary between the Massif Central crystalline basement and the south-eastern sedimentary basin of France. After several tectonic episodes that reactivated faults with different kinematics, this region is now characterized by a structural pattern with ~100 km long inherited faults, striking in NE-SW direction and dipping towards southeast. The La Rouvière fault lies between two main faults of the Cévennes fault zone, named as the Pontet-De-Couloubre fault and the Marsanne fault, as shown in Fig. 16. In the French Active Fault database (BFDA, Ref. [60]), the Marsanne Fault is interpreted as a potentially active left-lateral fault with a vertical slip component based on preliminary interpretations of offset geomorphic markers, whereas the age of Pontet-De-Couloubre fault's last activity is undetermined [60]. However, the La Rouvière fault was not identified as a potentially active fault in the same database before the earthquake.

Seismological analysis carried out after the 2019 earthquake indicated a reverse focal mechanism along the La Rouvière fault [61]. In addition, in situ stress measurements from boreholes, focal mechanism solutions, and geodetic observations show that the region is located within a NW–SE compressional regime [62–64]. This area has been struck by several shallow earthquakes with focal depths less than 5 km and macroseismic intensities up to VII (according to the Medvedev-Sponheuer-Karnik intensity scale), such as the August 1873, and May 1934 earthquakes reported in historical catalogues (e.g. Ref. [65]). It is notable that no surface ruptures were documented for these historical earthquakes.

Due to the well-recorded surface ruptures associated with such a small magnitude earthquake and its reverse kinematics, the Le Teil town that is located in the footwall of the La Rouvière fault has been selected for the IAEA PFDHA exercise (Fig. 16).

6.1. PRINCIPAL AND DISTRIBUTED FAULT DISPLACEMENT EXERCISES

One principal and three distributed displacement exercise cases were developed by considering the faults surrounding the Le Teil town, and a full set of seismic source parameters were provided to the fault displacement hazard analysts to explore the effect of the site location being on the hanging wall or footwall, and the dip angle on the results (Table 11). The site dimensions (z) were defined as 100 × 100 m for all (principal and distributed displacement) cases. The seismic source parameters summarized in Table 12 were defined by using the most recent and reliable information available in the literature and consulting the local experts. The final set of parameters were reviewed and approved by the Advisory Board.

In the principal fault displacement case, the site was assumed to be located on the principal fault (yellow star in Fig. 16), close to the middle of the segment with l/L=0.46. The epistemic uncertainty in fault length, fault width, dip angle, and mean annual rate of occurrence was discarded for simplicity. The hazard analysts were asked to consider a single magnitude value (M_W=5.5) and ignore the aleatory variability in the magnitude of the ruptures.

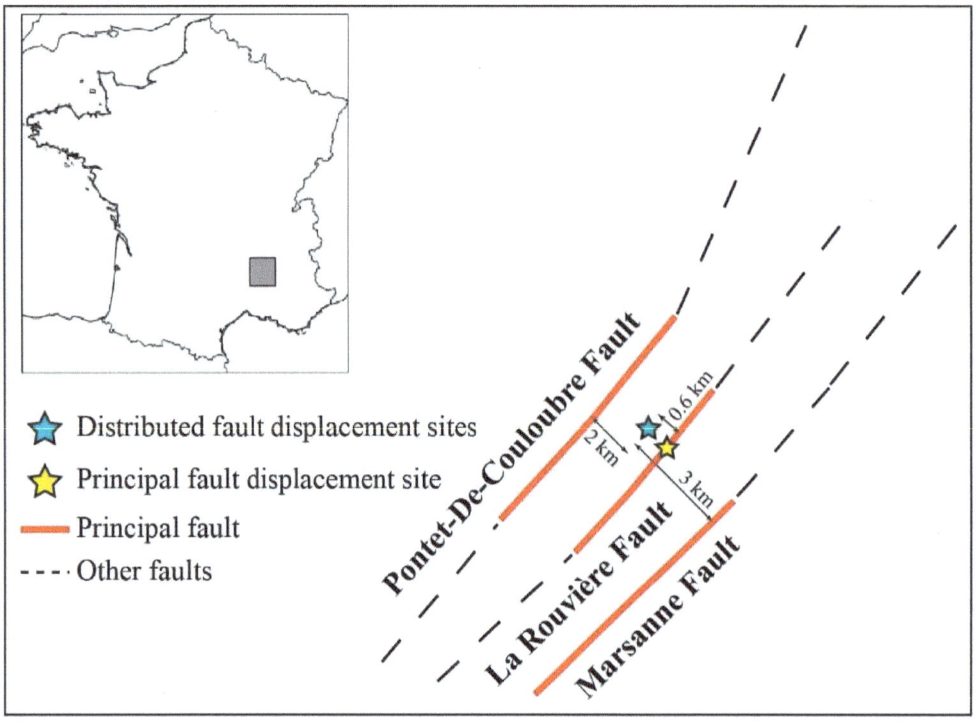

FIG. 16. Simplification of fault systems in the surrounding of Le Teil, and site locations for principal and distributed fault displacement exercises. The grey square in the insert shows the location of the site area.

TABLE 11. SUMMARY OF PARAMETERS USED FOR LE TEIL PRINCIPAL AND DISTRIBUTED FAULT DISPLACEMENT EXERCISES.

Cases	*r* (km)	Site Coordinates	Source	*l/L*	Map Accuracy
Principal	0	44.531, 4.67	La Rouvière Fault	0.46	-
1. Distributed	0.6	44.537, 4.667	La Rouvière Fault (FW)	0.32–0.46	Inferred
2. Distributed	0.6	44.537, 4.667	La Rouvière Fault (HW)	0.46	Inferred
3. Distributed	0.6	44.537, 4.667	La Rouvière Fault	0.46	Inferred
	1.9		Pontet-De-Couloubre Fault	0.27	
	2.9		Marsanne Fault	0.15	

A dip angle of 45° was assigned to the La Rouvière fault, based on the focal mechanism solutions of the 2019 earthquake sequence. It is notable that more precise seismological work recently concluded that the dip angle at depth (and at the surface) is 60°. Currently, these faults exhibit listric geometries and are rooted at depths of 3–5 km in relatively flat basal detachments within the Carboniferous-Triassic formations or early Jurassic strata [66, 67]. Therefore, the seismogenic depth for the La Rouvière Fault was assumed to be 4 km for the principal displacement exercise. An assumed seismogenic depth and dip angle were used in the magnitude-rupture area scaling relationships proposed by Ref. [68] to calculate the median rupture length as 5.8 km following a width-limited approach, which considers the concept of aspect ratio (rupture length per rupture width, *L/W*). The slip rate over the dip-slip surface was assumed as 0.01 mm/year and the annual mean rate of occurrence was computed as 4.6×10^{-5} a^{-1} by dividing the moment accumulation rate by the seismic moment per event (Tables 11, 12). The slip rate value was the best estimate provided by the local experts with the knowledge available at the time of source model development.

TABLE 12. SEISMIC SOURCES AND PARAMETERS USED FOR THE REVERSE CASE STUDY

Source	Maximum Rupture Length (km)	Maximum Rupture Thickness (km)	Average Dip (°)	Magnitude	Slip Rate (mm/a)
La Rouvière Fault	5.8	4	45 SE	5.5	0.01
La Rouvière Fault	5.8	4	45 NW	5.5	0.01
Marsanne Fault	6.5	6	80 SE	5.6	0.01
Pontet-De-Couloubre Fault	3.4	2	70 SE	4.9	0.01

Three distributed fault displacement cases were defined based on a single location the source-to-site distance of 0.6 km and l/L=0.46, as shown by the blue star in Fig. 16. The first distributed displacement exercise considers only the La Rouvière fault as the seismic source. For this case, the hazard analyst teams provided the PFDHA results for single path and multi path cases. In the single path case, the seismic source parameters of the principal fault displacement case were adopted without any change and the epistemic uncertainty in these parameters were ignored. Similar to the Kumamoto case, the centre branches of the logic tree for the multi path hazard runs includes the parameters of the single path case (Fig. 17). However, the logic tree implemented for the multi path case explores the uncertainties in the seismogenic thickness, rupture length, magnitude, and slip rate. It is notable that the range of uncertainties covered by the logic tree is quite large for some parameters, e.g. the seismogenic thickness is in the range 2–10 km, due to the shallow nature of the event and findings of the geophysical observations. This large range in the seismogenic thickness results in a large range of uncertainty in the magnitude, varying the magnitude between M_W=4.7 and M_W=6.7. On the other hand, the uncertainty in the slip rate was not that high for a case with limited knowledge on the actual slip rate. In addition, the Advisory Board asked for an additional branch in the multi path logic tree to account for the field observations after the 2019 event, which mapped the rupture length on the La Rouvière fault as 4.6 km (Fig. 17). This branch was added only to the path where the seismogenic thickness was set to 2 km.

The second distributed fault displacement exercise was similar to the first one with a single path, except that the dip direction of the La Rouvière fault was assigned as 45° NW (instead of 45° SE for the first case). This exercise was created to evaluate the impact of a potential change in dip direction from the south where the fault dip is well characterized to the north below Le Teil. This change also moves the test site from the footwall to the hanging wall of the fault.

The third distributed fault displacement exercise case considered the hazard given by the Pontet-De-Couloubre, La Rouvière, and Marsanne faults located in the surroundings of Le Teil town (Fig. 16). This exercise aimed to calculate the total hazard at the site and evaluate the contribution of each source to the total hazard, whilst testing the effect of source-to-site distances. Tables 11 and 12 present the parameters implemented in the third distributed fault displacement case.

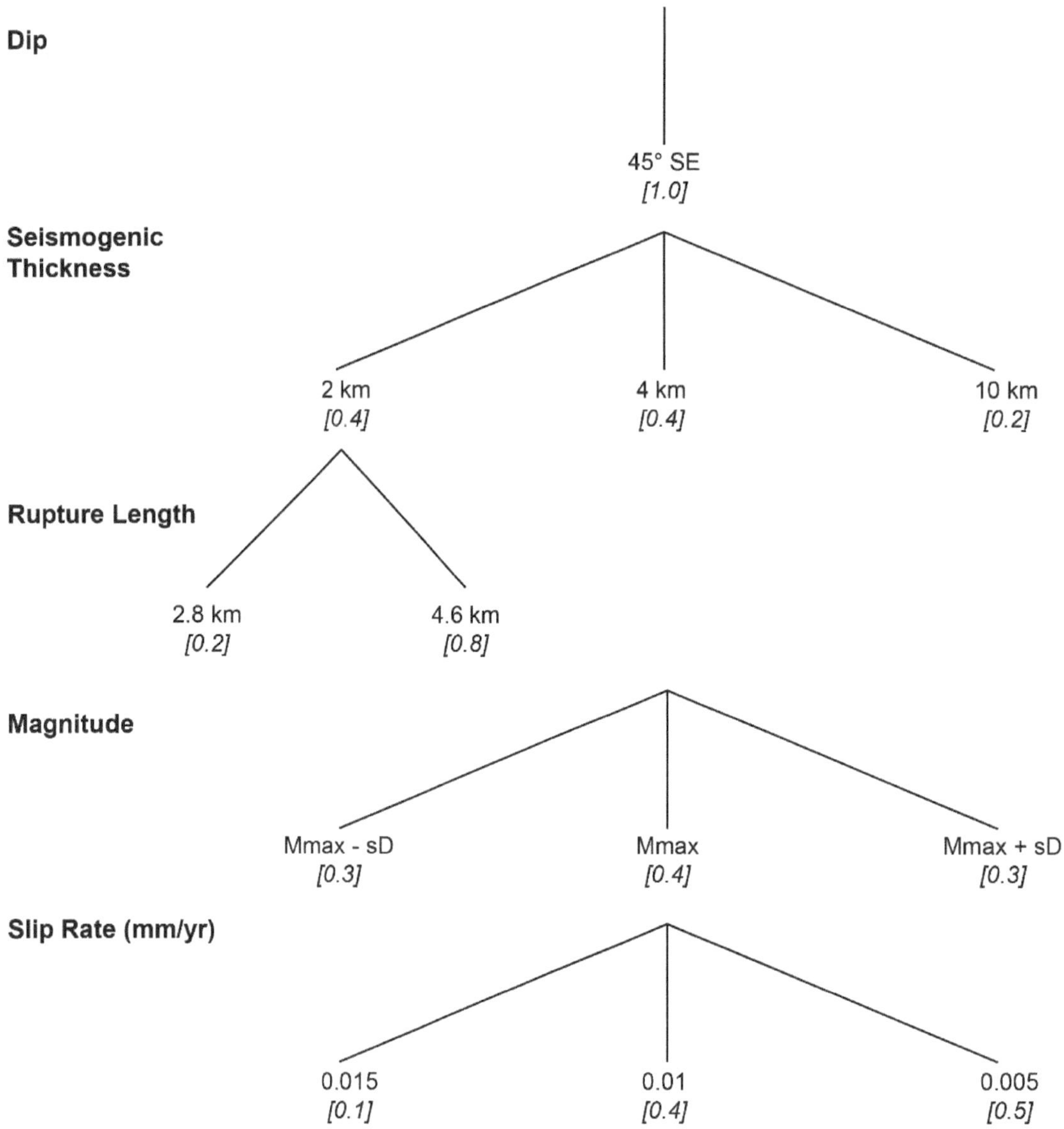

FIG. 17. Logic tree used in the multi path option of the first distributed fault displacement exercise. The weight of each branch is shown in square brackets. The magnitude of the centre branch is given in Table 12. The rupture length is modified only for case where the seismogenic thickness is 2 km.

6.2. COMPARISON OF THE HAZARD RESULTS

For the Le Teil PFDHA exercise, five hazard analyst teams provided PFDHA results. One team analysed both principal and distributed fault displacement cases, and four teams evaluated the PFDHA only for the principal fault displacement case. In addition, one of the teams provided the PFDHA results only for the distributed fault displacement exercises (see Table 8). Figures 18 and 19 present the hazard curves for the principal and distributed fault displacement exercises, respectively.

The difference in hazard curves for principal fault displacement exercise is remarkable (Fig. 18). The probability of surface rupture for a M_W=5.5 event is very small (~1%) in the T13 model proposed and the P_{2p} factor scaled down the hazard curve even further; therefore, the hazard estimates for that model are 10–100 times lower than the other models. One hazard analyst team (M11 in Fig. 18) combined the surface rupture probability model proposed by Ref. [30] (leading to 10% probability for M_W=5.5 earthquake) with the fault displacement model proposed by Ref. [23]. PFDHA based on the models provided by Refs [26, 28] resulted in the highest hazard estimates (L23, K24). This is because the team that implemented the K24 PFDHA model assumed that the probability of surface rupture is equal to 1, which is 10 times higher than the team that used the surface rupture probability model from Ref. [30]. The team that implemented the L23 model calculated the case-specific surface rupture probability by floating the rupture plane over the dipping surface of the fault within the hazard calculation. This resulted in a hazard curve that is close to the hazard curve of K24, which assumed that the probability of surface rupture is equal to 1. It is notable that the Le Teil earthquake is a very shallow earthquake with 4 km seismogenic depth. Therefore, the applicability of the empirical probability of surface rupture models shown in Fig. 4 may be debateable for this exercise case.

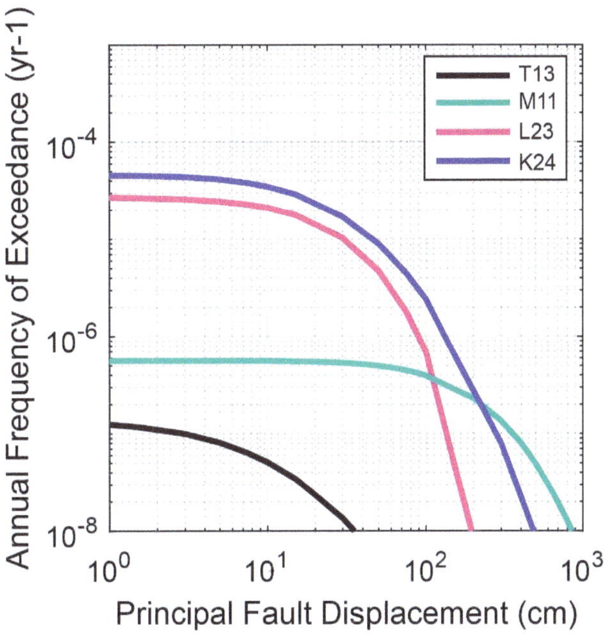

FIG. 18. Principal fault displacement hazard curves in terms of annual frequency of exceedance versus fault displacement for the principal fault displacement exercise. For model abbreviations, please refer to Table 8.

Only two models can predict distributed displacement for reverse earthquakes (T13 and V24). In the first distributed fault displacement exercise with a single path option (Fig. 19(a)), the T13 PFDHA model represents the upper bound for all displacement values. Nevertheless, the distributed fault displacement estimates are very small, reaching only to 10 cm at AFOE$\approx 10^{-10}$ a^{-1} for both models. Figure 19(b) shows the mean and median distributed displacement hazard curves, and 90% confidence intervals for the multi path option of the first distributed fault displacement exercise case. The 90% confidence intervals are quite broad, due to the large range of epistemic uncertainty in the logic tree given in Fig. 17. The 90% confidence intervals are about two orders of AFOE magnitude for the PFDHA runs implementing the model proposed by Ref. [22], and even larger for the runs with the model of Ref. [38].

The second distributed fault displacement exercise (Fig. 19(c)) uses the same seismic source of the first distributed exercise case with a different dip angle and consequently the site is located in the hanging wall instead of the footwall of the fault. Compared to Fig. 19(a), the hazard curve for the T13 model does not change, as this model does not include a separate term for the site location being on the hanging wall. The hazard curve for the model proposed by V24 increases compared to the hazard curve given in Fig. 19(a), indicating that the hanging wall scaling of the model proposed by Ref. [38] is notable.

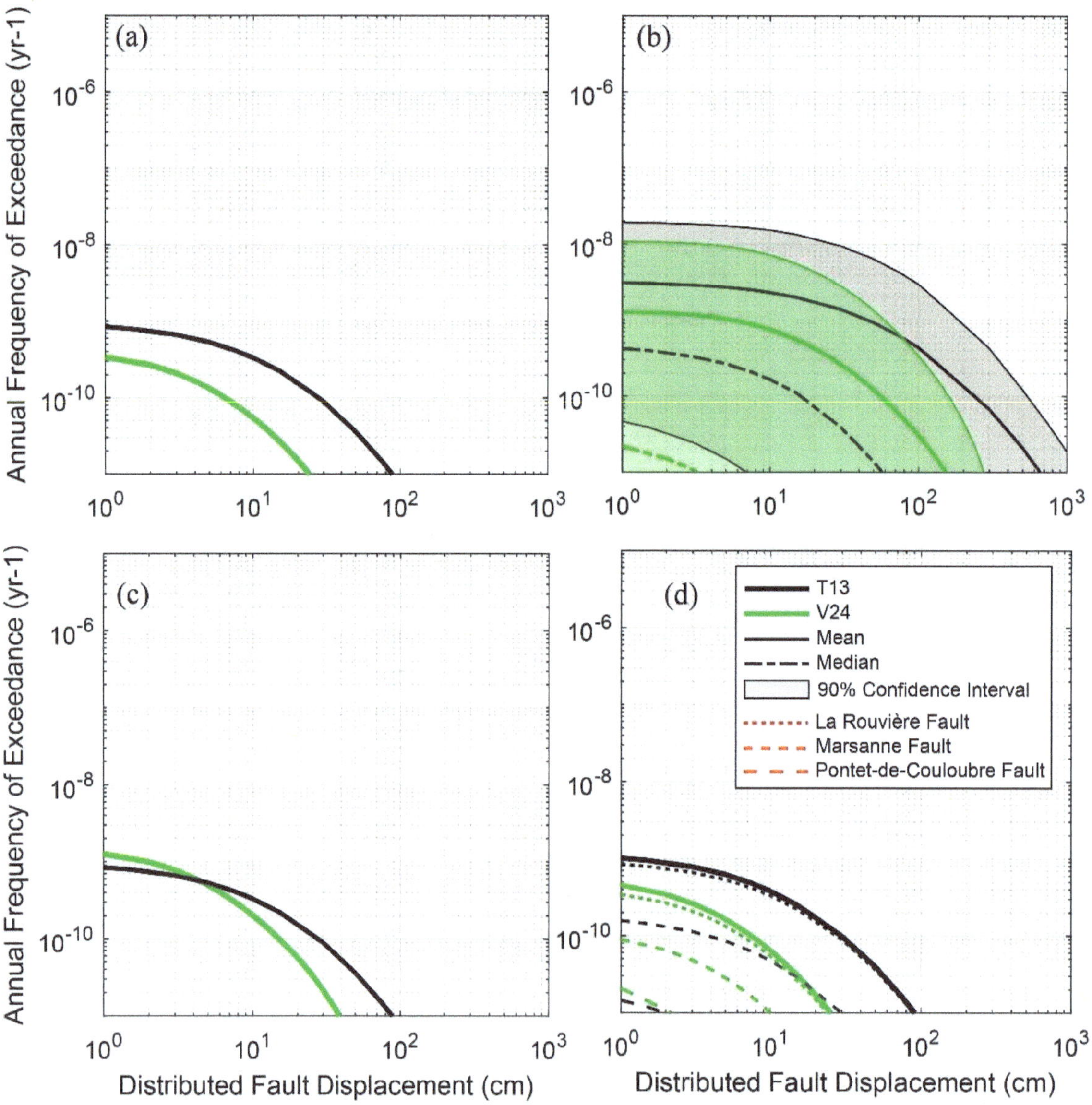

FIG. 19. Distributed displacement hazard curves in terms of annual frequency of exceedance (AFOE, a^{-1}) versus distributed fault displacement (cm). (a) 1st distributed displacement case, results of single path, (b) 1st distributed displacement case, results of multi path, (c) 2nd distributed displacement case (dip of the fault is changed to 45° NW), and (d) 3rd distributed displacement case (3 faults are included). For model abbreviations, please refer to Table 8.

The hazard curves for third distributed fault displacement exercise are shown in Fig. 19(d). This case accounts for three different seismic sources; therefore, the total and individual hazard curves are presented for both models. Independent of the model used to calculate the PFDHA, the La Rouvière fault is the dominating seismic source, followed by the Marsanne fault. The contribution of the Pontet-De-Couloubre fault to the total hazard is small because the site is located on the footwall side of the fault with source-to-site distance of 2 km.

The main conclusions and the lessons learned from the Le Teil exercise cases are discussed in Section 8.

7. THE NORMAL FAULTING CASE STUDY: NORCIA

The 2016 Central Italy earthquake sequence was one of the most destructive earthquake sequences that occurred in Italy in the last century. It reached the highest macroseismic intensity levels in Italy since the M_W=7.1 Fucino earthquake that occurred in 1915. It began on 24 August 2016, with a M_W=6.0 earthquake, followed by two moderate earthquakes on 26 October 2016 (M_W=5.4 and M_W=5.9), and culminated on 30 October 2016 with a M_W=6.5 earthquake. The mainshock struck ~6 km north of Norcia and the length of the surface rupture on the Mount Vettore fault was ~30 km.

The Mount Vettore fault, reactivated during the 2016 sequence, is located within the Apennines of central Italy, an area presently affected by NE-SW extension rates of 2–3 mm/a, according to Global Positioning System data [69]. The extension affected the central Apennines since Late Pliocene and is accommodated by a series of NW-SE striking normal fault systems, dipping mostly to the SW [70, 71], which led to the opening of Quaternary intermontane basins [72, 73]. The activity of these normal fault structures has been constrained by paleo earthquake, historical, and instrumental seismic records [74, 75, 76].

The site selected for the IAEA PFDHA exercise is located between the Mount Vettore and Norcia faults (Fig. 20). Both faults are considered active, with only the Mount Vettore fault slipping during the 2016 earthquake sequence. In addition to the evidence of displacement on the primary fault (maximum displacement of 2.4 m documented in Ref. [77]), secondary ruptures occurred as well. The largest displacement on secondary faults occurred on the pre-existing synthetic and antithetic splays in the hanging wall of the primary fault. The largest secondary rupture occurred on the 5 km long pre-existing San Lorenzo antithetic normal fault, which had a surface rupture of about 3.1 km in length with a maximum vertical displacement of 1.4 m. Another example of secondary rupture occurred on the 4 km long antithetic Mount Serra normal fault. It was responsible for the faulting of the 4.4 km long San Benedetto tunnel (National Road 685), located ~7 km from the primary fault, on the hanging wall, where a 20 cm vertical and a13 cm sinistral offset were measured [78].

Due to the well-recorded fault displacement both on the principal and secondary faults and the normal kinematics, a site between the Mount Vettore and Norcia faults (Fig. 20) was selected for the IAEA PFDHA exercise.

7.1. PRINCIPAL AND DISTRIBUTED FAULT DISPLACEMENT EXERCISES

The Norcia exercise included one principal and three distributed displacement cases to explore differences in the source-to-site distance and magnitude-frequency distribution (Table 13). The principal fault displacement exercise case site is located on the Mount Vettore fault trace, close to its southern termination (Fig. 20). Two distributed fault displacement exercise sites are located on the hanging wall of the Mount Vettore fault and on the footwall of the Norcia fault. The first site is located on the Mount Serra fault, at the point where the San Benedetto tunnel was displaced. The second is located on the trace of the San Lorenzo antithetic fault, at the point closest to the Mount Vettore fault trace. The site dimension (z) is the same for both the base and sensitivity cases and is 100 m × 100 m. The source parameter characterization (Table 14) for the principal and all distributed fault displacement sites of the Norcia case study were provided by the Advisory Board based on the most recent and reliable information available in the literature (e.g. Ref. [79]) as well as consultation with local experts.

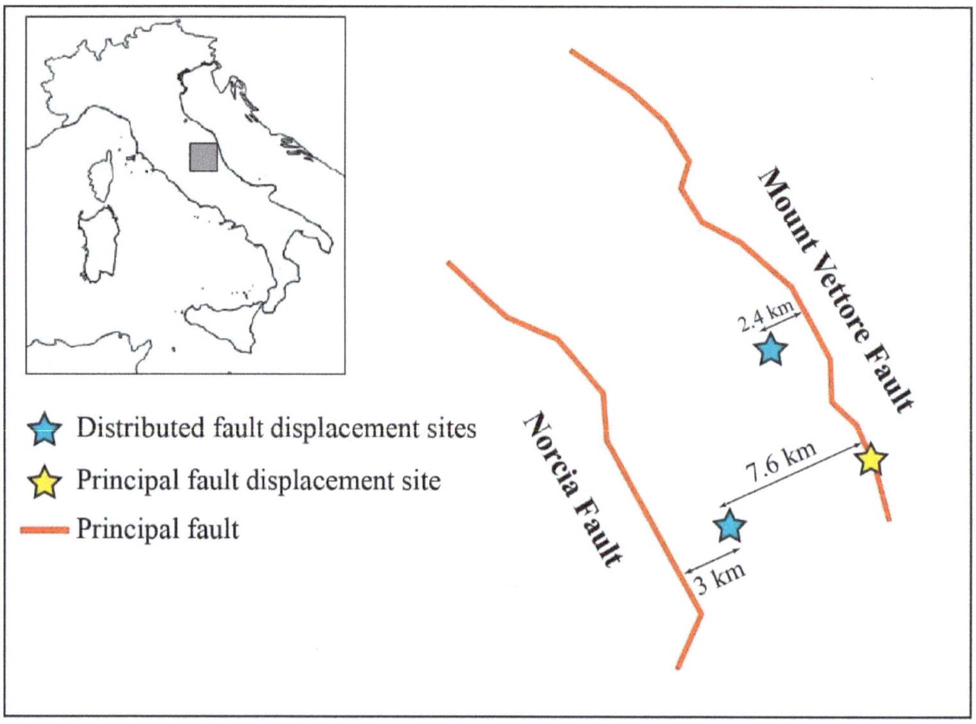

FIG. 20. Simplification of fault systems in the surrounding of Norcia, and site locations for principal and distributed fault displacement exercises for the normal case. The grey square in the insert is the location of the site area.

TABLE 13. SUMMARY OF PARAMETERS USED FOR NORCIA PRINCIPAL AND DISTRIBUTED FAULT DISPLACEMENT EXERCISES

Cases	*r* (km)	Site Coordinates	Source	l/L	Map Accuracy
Principal	0	42.767, 3.278	Mount Vettore Fault	0.05	-
1. Distributed	7.6	42.749,13.188	Mount Vettore Fault	0.05	Accurately located
2. Distributed	2.4	42.853,13.212	Mount Vettore Fault	0.4	Accurately located
3. Distributed	7.6 3	42.749,13.188	Mount Vettore Fault Norcia Fault	0.05 0.25	Accurately located

TABLE 14. SOURCES AND PARAMETERS USED FOR SEISMIC SOURCE CHARACTERIZION IN THE NORMAL CASE STUDY

Source	Maximum Rupture Length (km)	Maximum Rupture Thickness (km)	Average Dip (°)	Magnitude	Slip Rate (mm/a)	Mean Rate of Occurrence ($x10^{-4} a^{-1}$)
Mount Vettore Fault	34	11	47 SW	6.7 ± 0.3	0.75	4.03
Norcia Fault	29	12	50 SW	6.8	0.8	54.1

In the principal displacement case, the site was assumed to be located on the principal fault almost at the SE end of the Mount Vettore Fault (l/L=0.05), as shown by yellow star in Fig. 20. According to recent literature (e.g. Ref. [79]), each source is treated as individual seismogenic source, i.e. an individual structure liable to generate major earthquakes (e.g. $M_W \geq 5.5$) and where the rupture length and area are used to infer the slip and maximum expected magnitude. The geometrical fault parameters, such as length, dip, and seismogenic thickness, needed to infer the magnitude and mean seismic activity rate, were taken from the literature, and updated with information collected after the 2016 Central Italy earthquake sequence [79–83]. A 34 km long single segment fault dipping 47° towards SW was modelled. The seismogenic thickness was assumed to be 11 km based on the depth distribution of aftershocks [76, 84]. The slip rate used in conjunction with the geometric parameters to calculate the global budget of the seismic moment rate was taken as 0.75 mm/year [79]. Similar to the other cases, the epistemic uncertainty in these parameters was ignored for simplicity and the hazard analysts were asked to consider a single magnitude value (M_W=6.7). The mean annual rate of occurrence for M_W=6.7 earthquake was calculated as 40.3×10^{-5} a^{-1}.

The three distributed fault displacement cases were used to explore the sensitivity of hazard to magnitude-frequency distribution and r. For the first distributed fault displacement exercise, the Mount Vettore fault is considered as the only source of hazard. The site selected for the first distributed displacement exercise is situated where the San Benedetto Tunnel was displaced (blue star in Fig. 20). The point where the San Benedetto Tunnel was displaced is located at a distance of 7.6 km from the primary fault (hanging wall side) and experienced vertical and sinistral offsets of 20 cm and 13 cm, respectively [78]. Modellers provided hazard curves only for a single path case, where the epistemic uncertainty in source characterization is not explored. The source characterization is the same employed in the principal fault displacement exercise.

For the second distributed fault displacement case, the closest distance r between the Mount Vettore fault and the San Lorenzo antithetic fault is 2.4 km. In the first and second exercises, the faults were assumed to be accurately located and epistemic uncertainty in source parameters was not included. A third distributed fault displacement exercise was developed to compute the total hazard given by the contribution of all faults near the selected site (i.e. the Mount Vettore and Norcia faults). This case was designed to observe the contribution of each source to the hazard. The case also tests the different magnitude-frequency distribution and effect of the hanging wall vs. footwall. In addition to Mount Vettore fault, the Norcia Fault was also parameterized (Fig. 20). Its length was constrained at 29 km, the average dip angle is 50° SW, and the thickness of the seismogenic layer was constrained at 12 km. The geological slip rates, used with the geometric parameters to calculate the global budget of the seismic moment rate, was taken as 0.8 mm/a, the mean of the values determined by several authors in different ways over different time scales (e.g. from 10^3 to 10^5 years) and compiled in Ref. [80].

For the Norcia normal faulting case study, the Mount Vettore fault was modelled by a characteristic Gaussian model centred on the maximum magnitude and ranges from a lower bound to an upper bound (both defined by the maximum magnitude standard deviation) in accordance with Ref. [79]. The seismic moment rate was thus partitioned and balanced over this range of magnitudes. The Norcia fault was modelled by a Truncated Gutenberg-Richter model [85, 86], with a minimum moment magnitude threshold of 5.5 and the median + 1 standard deviation values were as the upper threshold. The b-value was taken as 1.0.

Tables 13 and 14 provide a summary of the cases implemented in this exercise and the parameters used.

7.2. COMPARISON OF THE HAZARD RESULTS

For the Norcia case study, four PFDHA teams provided fault displacement hazard curves considering the input information described in this section and provided by the Advisory Board. Only one team could provide results for the principal and all distributed fault displacement exercises. An additional team evaluated PFDHA only for the distributed fault displacement exercises and two further teams only analysed the principal fault displacement case. Figures 21 and 22 show the results for the principal and for the three distributed fault displacement exercises, respectively, as fault displacement hazard curves. The hazard curves are expressed in terms of annual frequency of exceedance (AFOE, a^{-1}) versus displacement (cm).

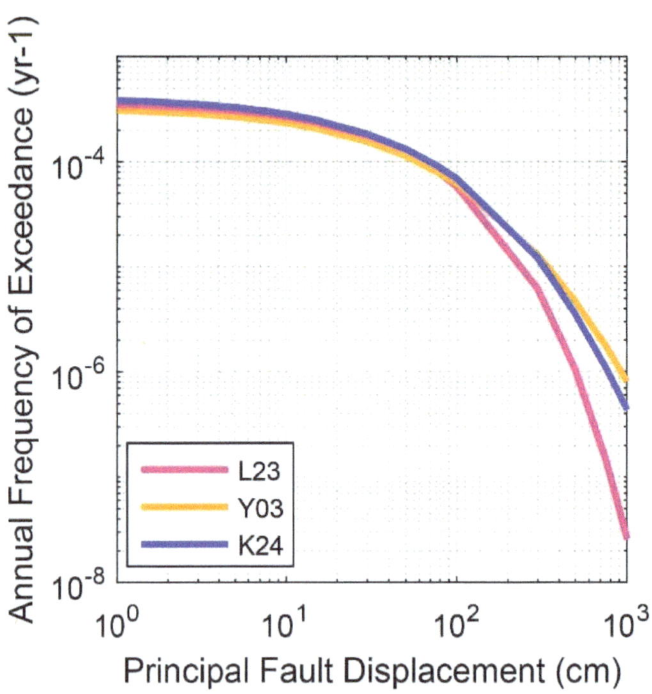

FIG. 21. Principal fault displacement hazard curves in terms of annual frequency of exceedance (AFOE) versus fault displacement for the principal fault displacement exercise. For model abbreviations, please refer to Table 8.

Figure 21 compares the principal fault displacement hazard curves for the Norcia exercise using the three PFDHA models listed in Table 8. The hazard curves are almost on top of each other up to a 10 000 year return period. Similar to the Kumamoto case, small differences in the hazard curves are associated with the differences in assumed surface rupture probabilities. There are significant differences in the hazard estimates for longer return periods, which may be explained by the aleatory variability models and truncation of the standard deviations. The three prediction models utilized in the Norcia exercise implemented three different probability distribution models for the standard deviation of the model. Similar to the ground motion models, the lognormal distribution was used by the hazard analysts that implement the Y03 model. However, because the tails of the distribution do not fit very well with the data, recent prediction models did not implement lognormal distribution. Power normal distribution was preferred by L23, and a Box-Cox transformation of the data was performed by K24, where the transformed data are more normally distributed. This distribution dominates the principal fault displacement estimates with 10^{-5} or lower annual probabilities. For the Norcia case, the principal fault displacement for a 100 000 year return period was estimated as 2 m by L23, 3.5 m by the Y03 model, and 5 m by the K24 PFDHA model. The hazard analysts were not provided a cut-off value to truncate the number of standard deviations above or below the median. Their choice in the value for truncating the aleatory variability model may have affected the results.

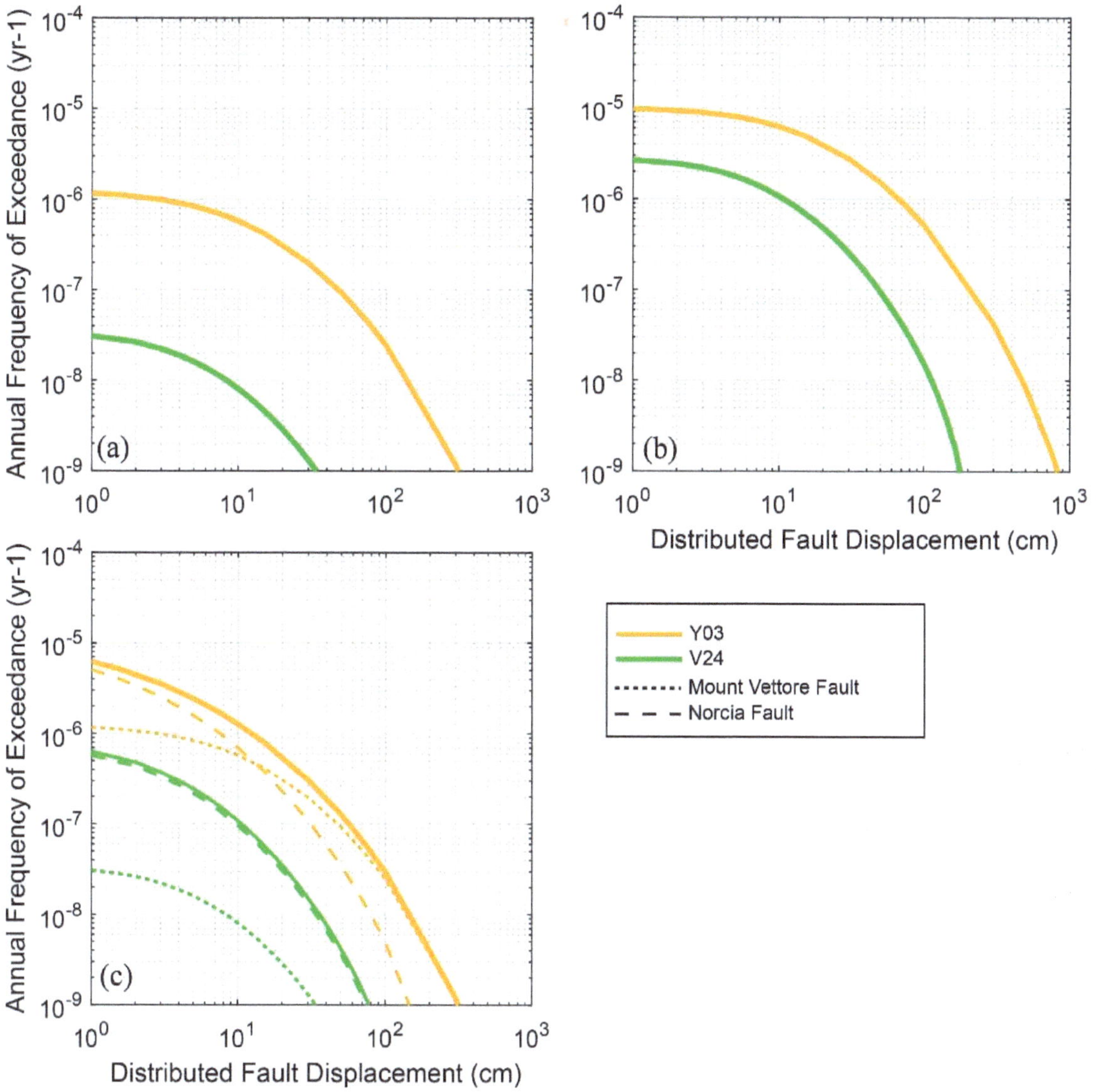

FIG. 22. Displacement hazard curves in terms of annual frequency of exceedance (AFOE, a^{-1}) versus displacement (cm), for the three distributed fault displacement exercises. (a) 1st distributed displacement case (r=7.6 km to Mount Vettore Fault), (b) 2nd distributed displacement case (r=2.4 km to Mount Vettore Fault), and (c) 3rd distributed displacement case (both faults are included). For model abbreviations, please refer to Table 8.

Only two models can predict distributed displacement for normal earthquakes. In the first distributed fault displacement exercise (Fig. 22(a)), the Y03 model represents the upper bound and V24 model results in displacements that are lower than in the Y03 model. The results represent the total hazard curve considering the Mount Vettore fault as the unique source of the hazard. Overall, the highest distributed displacements among all the distributed cases were estimated for the Norcia case, as this case has the highest mean annual rate of occurrence for the scenario earthquake (Table 14). The differences at the beginning of the hazard curves are related to the distributed rupture probability models shown in Fig. 9 (~ factor of 10). For AFOE=10^{-8} a^{-1}, the distributed displacement was estimated as ~115 cm by Y03 and 10 cm by V24 models. Moreover, the differences between the two hazard curves (Fig. 22(a)) are affected by the fact that in the V24 model the distributed rupture probability model is calibrated for the site dimension (100 × 100) and only the simple distributed rupture are considered (the so-called Rank 2 ruptures [38]). In contrast, the distributed probability model employed by Y03 is based on a grid of 500 × 500 and it consider all kinds of distributed rupture. Generally, if these two aspects are considered as in Y03, they lead to a more conservative hazard estimate.

The results of the second distributed fault displacement exercise are shown in Fig. 22(b). The input parameters are similar to the first distributed exercise with the exception of the source-to-site distance. A comparison between the first two distributed exercises shows that there is a consistent difference in the hazard results, and the hazard decreases when the source-to-site distance increases. Like the results of the first distributed exercise, the Y03 PFDHA model produces the highest distributed fault displacement hazard and V24 model produces the lowest hazard. For AFOE=10^{-6} a^{-1}, the distributed displacement was estimated as ~10 cm by V24 and 70 cm by Y03 models.

Figure 22(c) shows the hazard curves for the third distributed fault displacement exercise. This case accounts for the total hazard considering two different seismic sources. For the Y03 model, between 1 and 10 cm, the hazard is mainly due to the Norcia fault. In this case, the site is located in the footwall and 3 km away from the fault site. For displacement values greater than 10 cm, the hazard is mainly due to the Mount Vettore fault, which is 7.6 km away from the site located in the hanging wall. In the V24 model, the fault that contributes most to the total hazard is the Norcia fault, for all displacement values. Looking at the total hazard curves, for AFOE=10^{-7} a^{-1}, the distributed displacement was estimated as ~10 cm and 60 cm by the V24 and Y03 models, respectively.

The main conclusions and the lessons learned from the Norcia exercise cases are discussed in Section 8.

8. DISCUSSION OF THE FINDINGS, CONCLUSIONS AND LESSONS FROM THE EXERCISE

The IAEA PFDHA exercise provided valuable lessons on the important aspects of PFDHA, particularly on the selection of fault displacement and surface rupture models, logic tree development, and interpretation of the PFDHA results. This section presents a summary of these findings. The lessons from this interactive PFDHA exercise may also apply to other critical infrastructure such as dams, bridges, and lifelines.

8.1. DEVELOPMENT OF THE LOGIC TREE

One important limitation of the seismic source models used in the IAEA PFDHA exercise is the definition of the source model parameters in a deterministic manner, for easier realization of the underlying differences between the hazard curves provided by different teams. In PFDHA projects for nuclear installations and other critical infrastructure, the epistemic uncertainties in fault geometry, dip angles, activity rates, and maximum magnitude are included in the analysis using a seismic source characterization logic tree. In addition, the aleatory variability in earthquake magnitude is considered with appropriate magnitude probability density functions. As a result, it is harder for the decision makers to get an understanding of the differences in fault displacement estimations by different fault displacement models with such complex epistemic uncertainty structures.

The number of models for principal fault displacement has increased significantly in the last two to three years, owing to the Fault Displacement Hazard Initiative [87]. Therefore, developing a logic tree for fault displacement models, similar to the ground motion characterization logic tree of PSHA, is also possible for PFDHA. The IAEA PFDHA exercise demonstrated that the logic trees for seismic source characterization and fault displacement models may be separated to reduce the complexity of the application. The seismic source characterization logic tree will include the centre, body, and range for the technically defensible interpretations of the source model parameters, including fault geometry parameters, maximum magnitude, and activity rates. Variations in the rupture dimensions in the exercise underlined that the segmentation models need to be applied with caution and floating ruptures need be considered in addition to strict segmentation models to capture the aleatory variability in the rupture location and dimensions.

The logic tree structure for the principal fault displacement models will include branches for the surface rupture probability models, as shown in Fig. 23. The PFDHA exercise displayed that the principal fault displacement hazard curves are dominated by the applied surface rupture probability models for AFOE<10^{-4} a^{-1}, and empirical surface rupture probability models have a wide dispersion for M_W<6.5. Therefore, multiple surface rupture probability models may be selected by considering their applicability to the target area and alternative probability of surface rupture values may be included in the logic tree. The Le Teil exercise case pointed out the importance of this selection: even if the rupture depth of the Le Teil earthquake was very shallow and surface deformations were observed, empirical magnitude-surface rupture probability models lead to very small probabilities and significantly underestimate the probability of fault displacement for this case. Therefore, case-specific surface rupture probabilities may be calculated within the hazard analysis by moving the rupture plane over the dip-slip fault plane (as in PSHA) instead of using empirical surface rupture models. This alternative is indicated by the numerical branch in the logic tree given in Fig. 23.

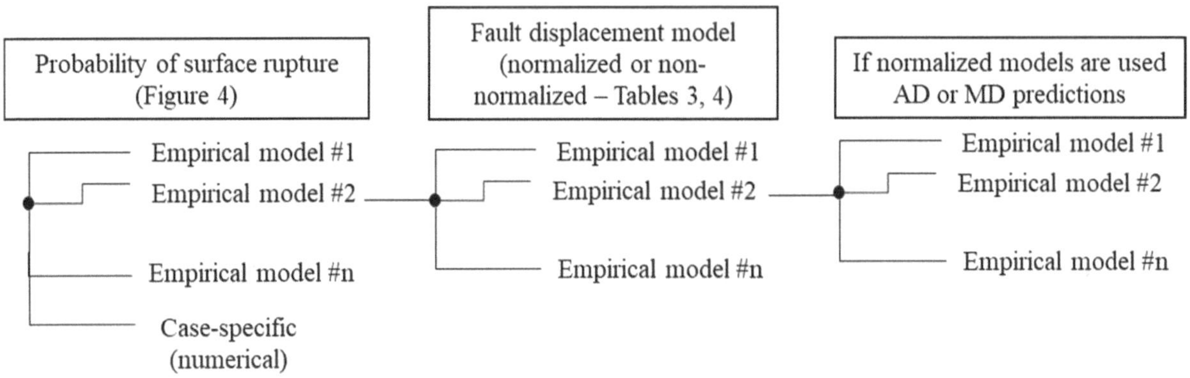

FIG. 23. Sketch of the fault displacement logic tree for principal PFDHA.

As the distributed fault displacements estimated for almost all exercise cases are quite low for AFOEs varying between 10^{-4} a^{-1} and 10^{-7} a^{-1}, screening of the fault displacement hazard may be possible by assessing the safety significance of the estimated displacements with preliminary analyses. The IAEA PFDHA exercise showed that the estimated distributed fault displacements are quite small, particularly for low-to-moderate magnitudes (i.e. $M_W<6$) and/or large source-to-site distances (r>2 km). In such cases, to demonstrate that screening of the hazard is possible, the upper bound models described in Sections 5–7 may be used in the preliminary analysis to get a conservative estimate of distributed fault displacement for AFOE = 10^{-7} a^{-1}. If this displacement is tolerable (i.e. it lies within the limits of structural reliability for structures important to safety), fault displacement logic trees may be simplified for distributed fault displacement. This issue is further discussed in Section 9. There are very few models for distributed fault displacement and most distributed fault displacement models had to be extrapolated to be applicable to small-to-moderate magnitudes and large source-to-site distances. Similar to PSHA, disaggregation plots may be checked to verify that the models are constrained in the magnitude range controlling the PFDHA.

The range of fractiles for the multi path case of the Kumamoto exercise is relatively small, mainly due to the simplified source characterization logic tree. In contrast, the range of fractiles for the multi path case of Le Teil is wider, as the uncertainties involved in some of the seismic source parameters are large. This is due to the difficulty in characterizing the seismic sources in low seismicity areas. If the PFDHA is to be conducted in such an environment, a major site characterization programme may be needed to reduce the epistemic uncertainty in the source parameters. Mean hazard curves cannot be solely relied upon, as they are influenced by extreme distribution tails. The difference between median and mean hazard curves can serve as an indicator of the level of epistemic uncertainty in complex logic trees, emphasizing the importance of considering the full range of fractiles.

Similar to the source models used in the IAEA PFDHA exercise, a preliminary and semi-deterministic seismic source model may be implemented at the beginning of the project to define the centre, body, and range of the logic tree for the fault displacement models, by allowing the experts and end users to make hazard-informed decisions. The treatment of uncertainties may be supported by sensitivity analyses to account for the limited knowledge and experience in PFDHA, when compared to PSHA. Plausibility assessments and 'sanity checks' are crucial to ensure consistency between results, input data, assumptions, and available observations.

The complexity of PFDHA models necessitates a detailed hazard input document that clearly describes all assumptions included in the hazard analysis. It is crucial to provide a thorough description of the choices made for each term of the PFDHA integral to avoid any misinterpretation. The numerical and graphical representation of the logic tree may help with the realization of independent calculations. The description of aleatory and epistemic uncertainties needs to be presented comprehensively to enable proper processing of the logic tree. If the standard deviations of any models are truncated (e.g. to the standard deviation of the magnitude scaling law), it needs to be clearly stated in order to allow reproducibility of calculations.

8.2. SELECTION OF THE FAULT DISPLACEMENT MODELS

Published fault displacement models are commonly developed for a specific style of faulting (also commonly referred to as sense of slip or fault kinematics). In some cases, the fault displacement model may be developed for a region that has a specific tectonic and kinematic environment, such as the model developed by Ref. [22] for regions of Japan with both reverse and strike-slip styles of faulting. Practitioners of PFDHA need to consider the tectonic environment and fault kinematics of the target area, compared to the tectonic environment and kinematics of earthquake ruptures used to constrain candidate fault displacement models in evaluating model applicability. In some instances, such as for cases of faults with an oblique sense of slip, practitioners may consider weighting models of more than one style of faulting.

The predicted fault displacement vector varies between models. Most models are developed for a specific style of faulting, and many models predict the dominant component associated with the style (e.g. [13, 14, 23, 33, 38, 39]). While some models predict the net component of displacement (e.g. [22, 24–28]), it is likely that measurements or estimates of net displacement in empirical datasets are systematically underestimated for dip-slip events because the fault-parallel (lateral) and fault-normal (heave) components are difficult to measure and seldom reported.

Practitioners need to consider differences in a model's fault displacement metric when comparing results from different models as this may be critical for different engineering applications or for using alternative models in a logic tree. For example, the model in Ref. [28] uses aggregate displacement, which is the sum of displacements across principal and distributed ruptures; the model in Ref. [27] only sums displacements across principal ruptures; and the model in Ref. [33] only considers single principal ruptures. The model in Ref. [26] provides two alternatives: an aggregate displacement model (comparable to Ref. [28]), and a summed principal model (comparable to Ref. [27]). Care is needed in interpreting the model results and applying the results to a site specific evaluation. Additionally, whether and/or how to adjust the hazard results to capture the net displacement vector may be important for the site or fault characteristics. In addition, adjustments for non-brittle deformation, such as warping or tilting, may also be needed because the currently available fault displacement models only predict discrete displacement.

Hazard curves need to be calculated across the range of surface displacements relevant to the engineering purpose. It is worthwhile noting that there is an upper bound on the surface displacements that can be physically generated by a finite fault, and these values correspond to the tails of the distribution. Establishing the range of displacement that the user is interested in needs to be agreed upon at the outset. When accounting for the aleatory variability in generating rupture scenarios, great care is needed to identify and discard non-realistic combinations. Employing moment balancing as a good practice in PSHA can help eliminate some unrealistic combinations.

The comparison of hazard curves between principal faulting and distributed faulting demonstrates the importance of accurate source characterization in PFDHA implementation. For NPP sites, the fault capability in the vicinity needs to be carefully assessed, even if they are minor faults. The influence of source-to-site distance on hazard curves appears to be relatively small, compared to other factors such as model selection. Further analyses, such as tornado plots and sensitivity analysis, need to be conducted to identify dominant uncertain factors and address the limited knowledge and experience in PFDHA. Model selection needs to consider the similarity of the tectonic environment and fault types to ensure appropriate representations. However, difficulties may arise when dealing with low-to-moderate seismicity contexts, as models may operate at the boundary of their validity domain.

9. A SEQUENTIAL SCREENING APPROACH FOR ASSESSING THE SAFETY SIGNIFICANCE OF PROBABILISTIC FAULT DISPLACEMENT HAZARD ANALYSIS RESULTS

Paragraph 7.10 of SSG-9 (Rev. 1) [2] states:

> "In the selection and evaluation stages of a proposed new site for a nuclear installation, if reliable evidence is collected demonstrating the existence of a capable fault with potential for seismogenic (i.e. primary) fault displacement within the site vicinity, or within the site area, and its effects cannot be compensated for by proven design or engineering protective measures, then the fault capability is an exclusionary attribute… and an alternative site should be considered."

Whether a fault can be considered as capable or not, in regard to its safety significance, is addressed in SSG-9 (Rev. 1) [2] and a detailed description of the overall process can be found in Ref. [88]. In some cases, however, the results of the PFDHA may be used for engineering analyses in order to support the safety assessments. The main objectives of this section are to provide practical information regarding the transposition of PFDHA results into engineering scenarios and to clarify the outputs of the PFDHA that will become inputs for safety assessment.

This section also emphases the necessity of developing a sequential screening approach for assessing the engineering implications of surface faulting in order to focus on safety significant issues and address any possible cliff edge effects.

In order to explore the centre, body and range of epistemic uncertainties in the PFDHA and to capture possible cliff edge effects, the approach for assessing the safety significance of surface faulting may rely on probabilistic considerations through fragility assessments of sensitive structures, systems and components (SSCs), instead of a deterministic approach that would not be informative enough [89, 90]. Based on the process in Ref. [88], key aspects of the sequential screening approach are as follows:

- Involving the full range of the PFDHA results (in terms of median hazard curves and other fractiles) as the input[2];
- Using the screening criteria in order to adapt engineering solutions and protective measures in multiple contexts (depending on the type of nuclear installation, site conditions, local seismotectonic context, national practices, etc.);
- Providing a fault displacement scenario that could be considered for engineering assessments (e.g. a simple and conservative scenario or more complex and realistic scenarios);
- Evaluating the fragility of sensitive SSCs in order to assess the safety significance of fault displacement hazard and possibly feed the probabilistic safety assessment (PSA) model as defined in Refs [91, 92][3].

---

[2] The PFDHA may be performed in stages, with the initial stage implementing a simplified but conservative logic tree, followed by a more complex logic tree if necessary (see Section 8).

[3] This step may also be iterative, with a simplified PSA model in the first stage followed by a full scope PSA model in a second stage, if necessary. This iterative process is not described in this publication.

Considering these key aspects, the main steps of this sequential screening approach are illustrated in Fig. 24. As shown by the multi path exercises for the Kumamoto and Le Teil cases, the epistemic uncertainty range in the PFDHA results may be quite large; therefore, a wide range of PFDHA curves needs to be considered in the sequential screening approach. The screening approach shown in Fig. 24 starts with a simple assessment of the magnitude of expected displacements. If the estimated fault displacements are less than a recognized lower bound capacity, the first level of screening may be applied.

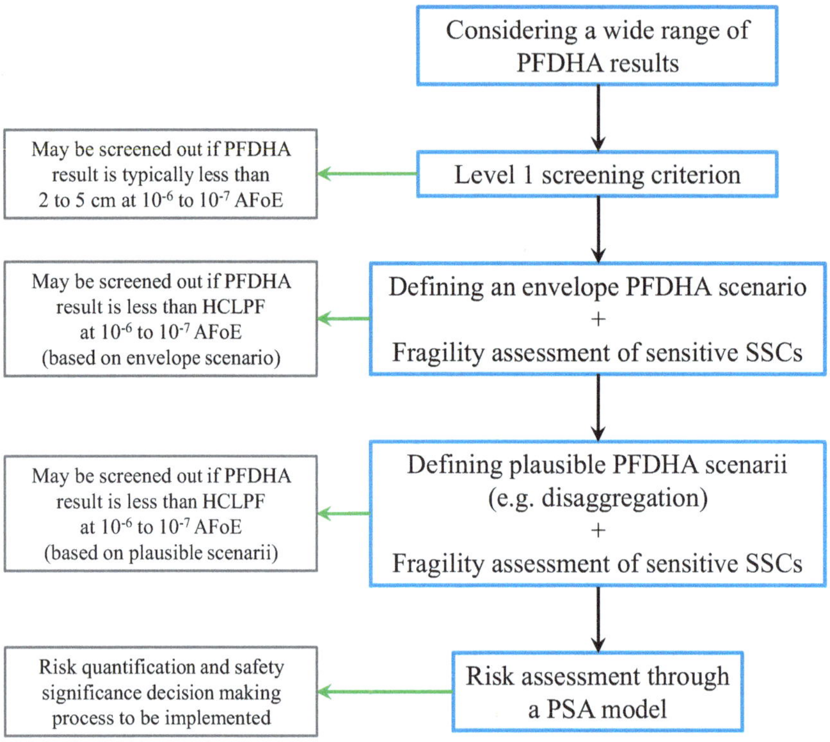

FIG. 24. Steps of the sequential screening approach for assessing the safety significance of fault displacement and probabilistic safety assessment for fault displacement hazard.

The screening concept described in Fig. 24 is in line with Requirement 6 of SSR-1 [1], which states that "**Potential external hazards... that could affect the region shall be identified through a screening process**". Based on the Member State experience of nuclear power plants in operation and published literature (e.g. Refs [88, 89]), a 2–5 cm fault displacement with 10^{-6} to 10^{-7} a^{-1} AFOE may be considered as a Level 1 high confidence low probability of failure (HCLPF)[4] value. It is widely believed that displacements of this magnitude will not lead to damage in engineered structures, considering that this level of displacement is usually close to other design displacement values for SSCs (e.g. long term settlement, seismic vibratory ground motion relative displacements, thermal expansion loads for piping systems, etc.). When coupled with low AFOEs of the order of 10^{-6} to 10^{-7} a^{-1}, these displacements would lead to a very low risk of non-acceptable consequences, so that the fault displacement hazard can be reasonably screened out. For the first level screening, other criteria may be adapted on a case-by-case basis evaluation of the nuclear installation conditions, in compliance with national practices regarding the general screening process for external hazards.

[4] High confidence low probability of failure is defined as a 5% probability of failure at a 95% confidence level (may sometime be considered as 1% probability of failure when mean fragility curves are used)

If the predicted displacements at 10^{-6} to 10^{-7} a^{-1} AFOE exceed the first screening criterion, the necessity of performing detailed engineering fragility assessments is confirmed and special care is needed in transforming the PFDHA results into engineering scenarios. This can be done either by defining an envelope scenario for simplified engineering applications or defining a scenario that correspond to plausible earthquake events (Fig. 24).

Defining an enveloping scenario for simplified engineering applications would consider one hazard curve (or a set of hazard curves including median and other fractiles) that provides a conservative estimation of the fault displacement, independent of the plausible scenario that could contribute to it. This approach is analogous to using the uniform hazard spectrum for the vibratory ground motions and may be implemented to demonstrate the absence of safety significance through a simple but conservative way. Defining a scenario (or multiple scenarios) that correspond to plausible earthquake events will consider multiple scenarios that could occur on the site, on the basis of the disaggregation of the hazard curve. This approach is analogous to using the conditional spectra or conditional mean spectra for vibratory ground motions and may be implemented to demonstrate the absence of safety issues through a more complex but realistic way by avoiding unnecessarily conservative assumptions.

For both options described above, analysts need to transpose the estimated ground displacements into a displacement vector that the structure, system, or component is subjected to. Transposing the PFDHA results into an imposed displacement field needs to account for:

- Fault rupture locations and orientations at the site (and corresponding uncertainties);
- Surface (or free field) hazard characterization, including the maximum and average displacement vectors in three dimensions and the spatial distribution displacements;
- In-depth hazard characterization including the maximum and average displacement vectors in three dimensions and the spatial distribution displacements if the engineering analyses intend to consider a detailed model for soil–structure interaction.

The engineering scenarios and displacement vectors may be provided for multiple AFOEs and/or fractiles to support the safety assessment. It is critical that the scenario and parameters are described in a format that would avoid any misunderstanding between the hazard assessment team and engineering assessment team. This description may consider uncertainties and could be complemented by simplified drawings, sketches, or any other necessary clarifications. It is encouraged that this step leads to a formal documented report that would be endorsed by both hazard assessment and engineering assessment teams.

At this stage of the sequential screening procedure (Fig. 24), the engineering scenarios will be used in fragility assessments for fault displacement. Fragility assessment is a well-established approach in seismic PSA for vibratory ground motion, relying on a large database of seismic experience data. The development of fragility functions for fault displacements is not as mature as the fragility functions for ground shaking (i.e. accelerations), however, there are some documented methodologies and applications (e.g. [89, 90]). In addition, fragility assessments for fault displacements may rely on quasi-static elastoplastic calculations. These are well-established practices, especially for mechanical components that might be sensitive to this kind of loading (e.g. differential displacements).

The outcome of this process would be a set of fragility curves providing the conditional probability of failure of SSCs, that may be used for screening or could feed a PSA model. Following the sequential screening approach given in Fig. 24, these set of fragility curves may also be derived into HCLPF values to feed (and possibly adapt) the second level screening criterion (HCLPF<10^{-6}–10^{-8}). The second level of screening may provide the opportunity to screen out the fault displacement hazard before launching a detailed PSA.

Finally, if the previous process does not allow to screen out the fault displacement safety significance, then the fragility of sensitive SSCs that have been established can be incorporated in a PSA model in order to explicitly quantify the risk and use the outcomes in a risk informed decision making process. At this stage, it might be necessary to consider the compliance of the evaluation of fault displacement and vibratory ground motion.

REFERENCES

[1] INTERNATIONAL ATOMIC ENERGY AGENCY, Site Evaluation for Nuclear Installations, IAEA Safety Standards Series No. SSR-1, IAEA, Vienna (2019).

[2] INTERNATIONAL ATOMIC ENERGY AGENCY, Seismic Hazards in Site Evaluation for Nuclear Installations, Specific Safety Guides SSG-9 (Rev. 1), IAEA, Vienna (2022).

[3] INTERNATIONAL ATOMIC ENERGY AGENCY, An Introduction to Probabilistic Fault Displacement Hazard Analysis in Site Evaluation for Existing Nuclear Installations, IAEA-TECDOC-1987, IAEA, Vienna (2021).

[4] NURMINEN, F. et al., SURE 2.0 - New release of the worldwide database of surface ruptures for fault displacement hazard analyses, Sci Data **9**(1) (2022) 729, https://doi.org/10.1038/s41597-022-01835-z.

[5] SARMIENTO, A. et al., Database for the Fault Displacement Hazard Initiative Project, Earthquake Spectra (2024), https://doi.org/10.1177/87552930241262766.

[6] SARMIENTO, A. et al., Comparisons of FDHI fault displacement models for aggregate and principal displacement (in preparation), Earthquake Spectra (2024), https://doi.org/10.34948/N3W88V.

[7] ROCKWELL, T.K. et al., Lateral Offsets on Surveyed Cultural Features Resulting from the 1999 İzmit and Düzce Earthquakes, Turkey, Bulletin of the Seismological Society of America **92**(1) (2002) 79, https://doi.org/10.1785/0120000809.

[8] MILLINER, C.W.D. et al., Quantifying near-field and off-fault deformation patterns of the 1992 Mw 7.3 Landers earthquake, Geochemistry, Geophysics, Geosystems **16**(5) (2015) 1577, https://doi.org/10.1002/2014GC005693.

[9] AATI, S., MILLINER, C., AVOUAC, J.-P., A new approach for 2-D and 3-D precise measurements of ground deformation from optimized registration and correlation of optical images and ICA-based filtering of image geometry artifacts, Remote Sensing of Environment **277** (2022) 113038, https://doi.org/10.1016/j.rse.2022.113038.

[10] SCOTT, C.P. et al., The M7 2016 Kumamoto, Japan, Earthquake: 3-D Deformation Along the Fault and Within the Damage Zone Constrained From Differential Lidar Topography, Journal of Geophysical Research: Solid Earth **123**(7) (2018) 6138, https://doi.org/10.1029/2018JB015581.

[11] JIANG, H., FENG, G., WANG, T., BÜRGMANN, R., Toward full exploitation of coherent and incoherent information in Sentinel-1 TOPS data for retrieving surface displacement: Application to the 2016 Kumamoto (Japan) earthquake, Geophysical Research Letters **44**(4) (2017) 1758, https://doi.org/10.1002/2016GL072253.

[12] WRIGHT, T.J., PARSONS, B.E., LU, Z., Toward mapping surface deformation in three dimensions using InSAR, Geophysical Research Letters **31**(1) (2004), https://doi.org/10.1029/2003GL018827.

[13] YOUNGS, R.R. et al., A Methodology for Probabilistic Fault Displacement Hazard Analysis (PFDHA), Earthquake Spectra, Vol. **19** (2003) pp. 191–219, https://doi.org/10.1193/1.1542891.

[14] PETERSEN, M.D. et al., Fault Displacement Hazard for Strike-Slip Faults, Bulletin of the Seismological Society of America **101**(2) (2011) 805, https://doi.org/10.1785/0120100035.

[15] BAIZE, S. et al., A Worldwide and Unified Database of Surface Ruptures (SURE) for Fault Displacement Hazard Analyses, Seismological Research Letters **91**(1) (2020) 499, https://doi.org/10.1785/0220190144.

[16] THOMPSON, J. et al., Evidence of Previous Faulting along the 2019 Ridgecrest, California, Earthquake Ruptures, Bulletin of the Seismological Society of America **110**(4) (2020) 1427, https://doi.org/10.1785/0120200041.

[17] SCOTT, C. et al., Evaluating how well active fault mapping predicts earthquake surface-rupture locations, Geosphere **19**(4) (2023) 1128, https://doi.org/10.1130/GES02611.1.

[18] KOTO, B., On the cause of the great earthquake in central Japan, 1891, Journal of College Sciences (Imperial University, Japan) **5**(4) (1893) 296.

[19] MCCALPIN, J.P., Paleoseismology, Academic Press (2009) 916 pp.

[20] WELLS, D., COPPERSMITH, K., Likelihood of surface rupture as a function of magnitude, Seismol. Res, Seismological Research Letters **64**(1) (1993) 54.

[21] PIZZA, M., FERRARIO, M.F., THOMAS, F., TRINGALI, G., LIVIO, F., Likelihood of Primary Surface Faulting: Updating of Empirical Regressions, Bulletin of the Seismological Society of America **113**(5) (2023) 2106, https://doi.org/10.1785/0120230019.

[22] TAKAO, M., TSUCHIYAMA, J., ANNAKA, T., KURITA, T., Application of Probabilistic Fault Displacement Hazard Analysis in Japan, Journal of Japan Association for Earthquake Engineering **13**(1) (2013) 17, https://doi.org/10.5610/jaee.13.17.

[23] MOSS, R.E.S., ROSS, Z.E., Probabilistic Fault Displacement Hazard Analysis for Reverse Faults, Bulletin of the Seismological Society of America **101**(4) (2011) 1542, https://doi.org/10.1785/0120100248.

[24] TAKAO, M. et al., Reliability Improvement of Probabilistic Fault Displacement Hazard Analysis, Journal of Japan Association for Earthquake Engineering **14**(2) (2014) 2_16, https://doi.org/10.5610/jaee.14.2_16.

[25] TAKAO, M., TANI, T., OSHIMA, T., ANNAKA, T., KURITA, T., Maximum Likelihood Estimation of the Parameters regarding Displacement Evaluation of Distributed Fault in PFDHA, Journal of Japan Association for Earthquake Engineering **16**(2) (2016) 2_96, https://doi.org/10.5610/jaee.16.2_96.

[26] LAVRENTIADIS, G., ABRAHAMSON, N., Fault-displacement models for aggregate and principal displacements, Earthquake Spectra (2023), https://doi.org/10.1177/87552930231201531.

[27] CHIOU, B.S.J. et al., Surface principal displacement model for strike-slip faults (under review), Earthquake Spectra (2024), https://doi.org/10.34948/N3RG6X.

[28] KUEHN, N., KOTTKE, A., SARMIENTO, A., MADUGO, C., BOZORGNIA, Y., A fault displacement model based on the FDHI database (under review), Earthquake Spectra (2024), https://doi.org/10.34948/N3X59H.

[29] CHEN, R., PETERSEN, M.D., Improved Implementation of Rupture Location Uncertainty in Fault Displacement Hazard Assessment, Bulletin of the Seismological Society of America **109**(5) (2019) 2132, https://doi.org/10.1785/0120180305.

[30] MOSS, R.E.S., STANTON, K.V., BUELNA, M.I., The Impact of Material Stiffness on the Likelihood of Fault Rupture Propagating to the Ground Surface, Seismol. Res. Lett. **84**(3) (2013) 485, https://doi.org/10.1785/0220110109.

[31] YANG, H., QUIGLEY, M., KING, T., Surface slip distributions and geometric complexity of intraplate reverse-faulting earthquakes, GSA Bulletin **133**(9–10) (2021) 1909, https://doi.org/10.1130/B35809.1.

[32] PEZZOPANE, S.K., DAWSON, T.E., Fault Displacement Hazard: A Summary of Issues and Information, in Seismotectonic Framework and Characterization of Faulting at Yucca Mountain, Nevada, U.S. Geological Survey Administrative Report prepared for the U.S. Department of Energy Chapter 9, (1996) p. 160.

[33] MOSS, R.E.S., THOMPSON, S.C., KUO, C.-H., YOUNESI, K., BAUMONT, D., Probabilistic Fault Displacement Hazard Analysis for Reverse Faulting (in press), Earthquake Spectra (2024), https://doi.org/10.34948/N3F595.

[34] WELLS, D.L., COPPERSMITH, K.J., New empirical relationships among magnitude, rupture length, rupture width, rupture area, and surface displacement, Bulletin of the Seismological Society of America **84**(4) (1994) 974.

[35] WHEELER, R.L., Persistent Segment Boundaries on Basin-Range Normal Faults, Palm Springs, California (1989) 432–444.

[36] MCCALPIN, J.P., SLEMMONS, D.B., STATISTICS OF PALEOSEISMIC DATA Program Element III: Understanding Earthquake Processes, Contract 1434-HQ-96-GR-02752, GEO-HAZ Consulting, Inc., Estes Park (1998).

[37] WESNOUSKY, S.G., Displacement and Geometrical Characteristics of Earthquake Surface Ruptures: Issues and Implications for Seismic-Hazard Analysis and the Process of Earthquake Rupture Displacement and Geometrical Characteristics of Earthquake Surface Ruptures, Bulletin of the Seismological Society of America **98**(4) (2008) 1609, https://doi.org/10.1785/0120070111.

[38] VISINI, F. et al., Empirical Regressions for Distributed Faulting of Dip-Slip Earthquakes (under review), Earthquake Spectra (2024).

[39] NURMINEN, F. et al., Probability of Occurrence and Displacement Regression of Distributed Surface Rupturing for Reverse Earthquakes, Front. Earth Sci. **8** (2020), https://doi.org/10.3389/feart.2020.581605.

[40] DEPOLO, C.M., The maximum background earthquake for the Basin and Range province, western North America, Bulletin of the Seismological Society of America **84**(2) (1994) 466, https://doi.org/10.1785/BSSA0840020466.

[41] FERRARIO, M.F., LIVIO, F., Conditional probability of distributed surface rupturing during normal-faulting earthquakes, Solid Earth **12**(5) (2021) 1197, https://doi.org/10.5194/se-12-1197-2021.

[42] VALENTINI, A. et al., Probabilistic Fault Displacement Hazard Assessment (PFDHA) for Nuclear Installations According to IAEA Safety Standards, Bulletin of the Seismological Society of America **111**(5) (2021) 2661, https://doi.org/10.1785/0120210083.

[43] VALENTINI, A., FUKUSHIMA, Y., CONTRI, P., GULERCE, Z., The IAEA Exercise on Probabilistic Fault Displacement Hazard Assessment (under review), Earthquake Spectra (2024).

[44] Webinar on Recent Advances in Probabilistic Fault Displacement Hazard Assessment for Nuclear Installations in Light of Geological Reconnaissance Findings, https://www.iaea.org/resources/webinar/webinar-on-recent-advances-in-probabilistic-

fault-displacement-hazard-assessment-for-nuclear-installations-in-light-of-geological-reconnaissance-findings.

[45] SHIRAHAMA, Y. et al., Characteristics of the surface ruptures associated with the 2016 Kumamoto earthquake sequence, central Kyushu, Japan, Earth, Planets and Space **68**(1) (2016) 191, https://doi.org/10.1186/s40623-016-0559-1.

[46] THE HEADQUARTERS FOR EARTHQUAKE RESEARCH PROMOTION (HERP), Long-term evaluation of the Futagawa and Hinagu fault zones (partly revised), (2013).

[47] ASANO, K., IWATA, T., Source rupture processes of the foreshock and mainshock in the 2016 Kumamoto earthquake sequence estimated from the kinematic waveform inversion of strong motion data, Earth, Planets and Space **68**(1) (2016) 147, https://doi.org/10.1186/s40623-016-0519-9.

[48] HIMEMATSU, Y., FURUYA, M., Fault source model for the 2016 Kumamoto earthquake sequence based on ALOS-2/PALSAR-2 pixel-offset data: evidence for dynamic slip partitioning, Earth, Planets and Space **68**(1) (2016) 169, https://doi.org/10.1186/s40623-016-0545-7.

[49] KOBAYASHI, T., Earthquake rupture properties of the 2016 Kumamoto earthquake foreshocks (Mj 6.5 and Mj 6.4) revealed by conventional and multiple-aperture InSAR, Earth, Planets and Space **69**(1) (2017) 7, https://doi.org/10.1186/s40623-016-0594-y.

[50] KOBAYASHI, T. et al., Crustal Deformation and Fault Models of the 2016 Kumamoto Earthquake Sequence: Foreshocks and Main Shock, International Symposium on Advancing Geodesy in a Changing World (FREYMUELLER, J.T., SÁNCHEZ, L., Eds), Springer International Publishing, Cham (2019) 193–200.

[51] FUJIWARA, S. et al., Small-displacement linear surface ruptures of the 2016 Kumamoto earthquake sequence detected by ALOS-2 SAR interferometry, Earth Planets Space **68**(1) (2016) 160, https://doi.org/10.1186/s40623-016-0534-x.

[52] GOTO, H., TSUTSUMI, H., TODA, S., KUMAHARA, Y., Geomorphic features of surface ruptures associated with the 2016 Kumamoto earthquake in and around the downtown of Kumamoto City, and implications on triggered slip along active faults, Earth, Planets and Space **69**(1) (2017) 26, https://doi.org/10.1186/s40623-017-0603-9.

[53] THE HEADQUARTERS FOR EARTHQUAKE RESEARCH PROMOTION (HERP), Evaluation of the Futagawa and Hinagu fault zones, (2016).

[54] MATSUDA, T., Magnitude and recurrence interval of earthquakes from a fault, Zisin 28, 269–284., Zisin (Journal of the Seismological Society of Japan. 2nd ser.) **28** (1975) 269, https://doi.org/10.4294/zisin1948.28.3_269.

[55] TAKEMURA, M., Magnitude-Seismic Moment Relations for the Shallow Earthquakes in and around Japan, Zisin (Journal of the Seismological Society of Japan. 2nd ser.) **43**(2) (1990) 257, https://doi.org/10.4294/zisin1948.43.2_257.

[56] MATSUDA, T., YAMAZAKI, H., NAKATA, T., IMAIZUMI, T., Earthquake fault of the Rikuu earthquake of 1896, Bull. Earthq. Res. Inst. Univ. Tokyo **55** (1980) 795.

[57] GAGNEPAIN, J. et al., Sismicité de la region d'Arette (P.A.) et mécanismes au foyer, Ann. Geophys. **36** (1980) 499.

[58] CARA, M., ALASSET, P.-J., SIRA, C., "Magnitude of Historical Earthquakes, from Macroseismic Data to Seismic Waveform Modelling: Application to the Pyrenees and a 1905 Earthquake in the Alps", Historical Seismology: Interdisciplinary Studies of Past

and Recent Earthquakes (FRÉCHET, J., MEGHRAOUI, M., STUCCHI, M., Eds), Springer Netherlands, Dordrecht (2008) 369–384, https://doi.org/10.1007/978-1-4020-8222-1_18.

[59] RITZ, J.-F. et al., Surface rupture and shallow fault reactivation during the 2019 Mw 4.9 Le Teil earthquake, France, Communications Earth & Environment **1**(1) (2020) 1, https://doi.org/10.1038/s43247-020-0012-z.

[60] JOMARD, H. et al., Transposing an active fault database into a seismic hazard fault model for nuclear facilities – Part 1: Building a database of potentially active faults (BDFA) for metropolitan France, Natural Hazards and Earth System Sciences **17**(9) (2017) 1573, https://doi.org/10.5194/nhess-17-1573-2017.

[61] DELOUIS, B. et al., Constraining the point source parameters of the 11 November 2019 Mw 4.9 Le Teil earthquake using multiple relocation approaches, first motion and full waveform inversions, Comptes Rendus. Géoscience **353**(S1) (2021) 493, https://doi.org/10.5802/crgeos.78.

[62] MASSON, C., MAZZOTTI, S., VERNANT, P., DOERFLINGER, E., Extracting small deformation beyond individual station precision from dense Global Navigation Satellite System (GNSS) networks in France and western Europe, Solid Earth **10**(6) (2019) 1905, https://doi.org/10.5194/se-10-1905-2019.

[63] BLES J., GROS, Stress field changes in the Rhone Valley from the Miocene to the present., Tectonophysics **194**(3) (1991) 265, https://doi.org/10.1016/0040-1951(91)90264-S.

[64] REBAÏ, S., PHILIP, H., TABOADA, A., Modern tectonic stress field in the Mediterranean region: evidence for variation in stress directions at different scales, Geophysical Journal International **110**(1) (1992) 106, https://doi.org/10.1111/j.1365-246X.1992.tb00717.x.

[65] MANCHUEL, K. et al., The French seismic CATalogue (FCAT-17), Bull Earthquake Eng **16**(6) (2018) 2227, https://doi.org/10.1007/s10518-017-0236-1.

[66] ROURE, F., BRUN, J.-P., COLLETTA, B., VAN DEN DRIESSCHE, J., Geometry and kinematics of extensional structures in the alpine foreland basin of southeastern France, Journal of Structural Geology **14**(5) (1992) 503, https://doi.org/10.1016/0191-8141(92)90153-N.

[67] THOMASSET, C., RITZ, J.-F., POULIQUEN, S., MANCHUEL, K., LE-ROUX-MALLOUF, R., Geometry and tectonic history of the Northeastern Cevennes Fault System (Southeast Basin, France): new insights from deep seismic reflection profiles, BSGF (2024), https://doi.org/10.1051/bsgf/2024016.

[68] LEONARD, M., Earthquake Fault Scaling: Self-Consistent Relating of Rupture Length, Width, Average Displacement, and Moment Release, Bulletin of the Seismological Society of America **100**(5A) (2010) 1971, https://doi.org/10.1785/0120090189.

[69] D'AGOSTINO, N., Complete seismic release of tectonic strain and earthquake recurrence in the Apennines (Italy), Geophysical Research Letters **41**(4) (2014) 1155, https://doi.org/10.1002/2014GL059230.

[70] BONCIO, P., LAVECCHIA, G., PACE, B., Defining a model of 3D seismogenic sources for Seismic Hazard Assessment applications: The case of central Apennines (Italy), Journal of Seismology **8**(3) (2004) 407, https://doi.org/10.1023/B:JOSE.0000038449.78801.05.

[71] CALAMITA, F. et al., Quaternary faults and seismicity in the Umbro-Marchean Apennines (Central Italy): evidence from the 1997 Colfiorito earthquake, Journal of Geodynamics **29**(3) (2000) 245, https://doi.org/10.1016/S0264-3707(99)00054-X.

[72] ROBERTS, G.P., MICHETTI, A.M., Spatial and temporal variations in growth rates along active normal fault systems: an example from The Lazio–Abruzzo Apennines, central Italy, Journal of Structural Geology **26**(2) (2004) 339, https://doi.org/10.1016/S0191-8141(03)00103-2.

[73] GALADINI, F., GALLI, P., Active Tectonics in the Central Apennines (Italy) – Input Data for Seismic Hazard Assessment, Natural Hazards: Journal of the International Society for the Prevention and Mitigation of Natural Hazards **22**(3) (2000) 225, https://doi.org/10.1023/A:1008149531980.

[74] GALLI, P., GALADINI, F., PANTOSTI, D., Twenty years of paleoseismology in Italy, Earth-Science Reviews **88**(1) (2008) 89, https://doi.org/10.1016/j.earscirev.2008.01.001.

[75] ROVIDA, A., LOCATI, M., CAMASSI, R., LOLLI, B., GASPERINI, P., The Italian earthquake catalogue CPTI15, Bulletin of Earthquake Engineering **18**(7) (2020) 2953, https://doi.org/10.1007/s10518-020-00818-y.

[76] CHIARALUCE, L. et al., The 2016 Central Italy Seismic Sequence: A First Look at the Mainshocks, Aftershocks, and Source Models, Seismological Research Letters **88**(3) (2017) 757, https://doi.org/10.1785/0220160221.

[77] BROZZETTI, F. et al., High-Resolution Field Mapping and Analysis of the August–October 2016 Coseismic Surface Faulting (Central Italy Earthquakes): Slip Distribution, Parameterization, and Comparison With Global Earthquakes, Tectonics **38**(2) (2019) 417, https://doi.org/10.1029/2018TC005305.

[78] GALLI, P., GALDERISI, A., MARTINO, M., SCARASCIA MUGNOZZA, G., BOZZANO, F., "The coseismic faulting of the San Benedetto tunnel (2016, Mw 6.6 central Italy earthquake)", Tunnels and Underground Cities: Engineering and Innovation Meet Archaeology, Architecture and Art, CRC Press (2020).

[79] VALENTINI, A. et al., Definition of Seismic Input From Fault-Based PSHA: Remarks After the 2016 Central Italy Earthquake Sequence, Tectonics **38**(2) (2019) 595, https://doi.org/10.1029/2018TC005086.

[80] VALENTINI, A., VISINI, F., PACE, B., Integrating faults and past earthquakes into a probabilistic seismic hazard model for peninsular Italy, Natural Hazards and Earth System Sciences **17**(11) (2017) 2017, https://doi.org/10.5194/nhess-17-2017-2017.

[81] CIVICO, R. et al., Surface ruptures following the 30 October 2016 Mw 6.5 Norcia earthquake, central Italy, Journal of Maps **14**(2) (2018) 151, https://doi.org/10.1080/17445647.2018.1441756.

[82] VILLANI, F. et al., Surface Faulting of the 30 October 2016 Mw 6.5 Central Italy Earthquake: Detailed Analysis of a Complex Coseismic Rupture, Tectonics **37**(10) (2018) 3378, https://doi.org/10.1029/2018TC005175.

[83] GALLI, P. et al., The Awakening of the Dormant Mount Vettore Fault (2016 Central Italy Earthquake, M 6.6): Paleoseismic Clues on Its Millennial Silences, Tectonics **38**(2) (2019) 687, https://doi.org/10.1029/2018TC005326.

[84] VALOROSO, L. et al., Radiography of a normal fault system by 64,000 high-precision earthquake locations: The 2009 L'Aquila (central Italy) case study, Journal of

Geophysical Research: Solid Earth **118**(3) (2013) 1156, https://doi.org/10.1002/jgrb.50130.

[85] ORDAZ, M., REYES, C., Earthquake hazard in Mexico City: Observations versus computations, Bulletin of the Seismological Society of America **89**(5) (1999) 1379, https://doi.org/10.1785/BSSA0890051379.

[86] KAGAN, Y.Y., Seismic moment distribution revisited: I. Statistical results, Geophysical Journal International **148**(3) (2002) 520, https://doi.org/10.1046/j.1365-246x.2002.01594.x.

[87] BOZORGNIA, Y. et al., An Overview of the Fault Displacement Hazard Initiative Research Program., Seismological Society of America 2021 Annual Meeting, (2021).

[88] GÜRPINAR, A., SERVA, L., LIVIO, F., RIZZO, P.C., Earthquake-induced crustal deformation and consequences for fault displacement hazard analysis of nuclear power plants, Nuclear Engineering and Design **311** (2017) 69, https://doi.org/10.1016/j.nucengdes.2016.11.007.

[89] U.S. NUCLEAR REGULATORY COMMISSION, Confirmatory Analysis of Seismic Hazard at the Diablo Canyon Power Plant from the Shoreline Fault Zone, Research Information Letters – 2012 ML121230035, (2012).

[90] HARAGUCHI, R. et al., Development of fault displacement PRA methodology and its application to a hypothetical NPP, Nuclear Engineering and Design **361** (2020) 110433, https://doi.org/10.1016/j.nucengdes.2019.110433.

[91] INTERNATIONAL ATOMIC ENERGY AGENCY, Development and Application of Level 1 Probabilistic Safety Assessment for Nuclear Power Plants, IAEA Safety Standards Series No. SSG 3(Rev.1), IAEA, Vienna (2024).

[92] INTERNATIONAL ATOMIC ENERGY AGENCY, Development and Application of Level 2 Probabilistic Safety Assessment for Nuclear Power Plants, IAEA Safety Standards Series No. SSG-4, IAEA, Vienna (2010). (a revision of this publication is in preparation).

ANNEX: USE OF THE POTENTIALLY DEFORMED AREA TO INFER THE TIME FRAME FOR FAULT CAPABILITY ASSESSMENT

Paragraph 7.3 of IAEA Safety Standards Series No. SSG-9 (Rev. 1), Seismic Hazards in Site Evaluation for Nuclear Installations [A–1] states:

> "The first question in assessing the potential for fault displacement is whether a fault (buried or outcropping) within the site vicinity and/or within the site area is to be considered capable (i.e., whether the fault has a significant potential for producing displacement at or near the ground surface). The basis for answering such a question should be the proper analysis and interpretation of the data compiled in the integrated database…, as incorporated in the seismic source models…, together with additional specific data that may be needed for such an assessment."

Therefore, the starting point of fault displacement hazard analysis for nuclear installations is the assessment of fault capability, whose definition strictly depends on the age of the last movement that occurred along a fault. Paragraph 7.4 of SSG-9 (Rev. 1) [A–1] states:

> "On the basis of the geological, geophysical, geodetic and/or seismological data, a fault should be considered capable if the following conditions apply:
>
> (a) If the fault shows evidence of past movement (e.g., significant deformations and/or dislocations) within such a period that it is reasonable to conclude that further movements at or near the surface might occur over the lifetime of the site or the nuclear installation, the fault should be considered capable. In highly active areas, where both seismic and geological data consistently reveal short earthquake recurrence intervals, evidence of past movements in the Upper Pleistocene to the Holocene (i.e., the present) might be appropriate for the assessment of capable faults. In less active areas, it is likely that much longer periods (e.g. the Pliocene to the Holocene (i.e. the present) are appropriate. In areas where the observed activity is between these two rates (i.e., not as highly active as plate boundaries and not as stable as cratonic zones), the length of the period to be considered should be chosen on a conservative basis (e.g. the Quaternary with possible extension to the Pliocene, depending on the area's tectonic activity level). One way to calibrate the time frame for fault capability would be to check whether the site is in the deformed area of major regional faults. Longer time frames should be used when the site is far away from the potentially deformed areas of these regional structures."

This underlines the need for calibrating the time frame that will be considered to assess the fault capability. SSG-9 (Rev. 1) [A–1] emphasize that the site specific time frame to be used in fault capability assessment needs to be defined at the beginning of site evaluation. In particular, para. 7.5 of SSG-9 (Rev. 1) [A–1] states:

> "The period within which evidence of past movement will determine the capability of a fault…should be defined at the beginning of the seismic hazard assessment project through a site-specific criterion based on the characteristics of the regional tectonic environment and the conditions in the near region and site vicinity. This criterion for assessing fault capability should be established by or agreed with the regulatory body."

The site specific time frame depends strongly on the seismotectonic characteristics of the region as noted above. 'Highly active areas' are those characterized by strong and frequent seismicity that are generally related to: (a) major plate boundaries like the so-called Ring of Fire or Eastern

Mediterranean arc; (b) ongoing orogenesis such as the Alpine chain from the Atlas to the Zagros or the Basins and Ranges; (c) new Rifts such as the African rift or the Rhine graben. The time frame for these areas is typically Upper Pleistocene to the Holocene (i.e. the present). In 'less active areas' (e.g. major cratons as in Central-Eastern Europe, Eastern US and Canada, Australia, South American Shield, Western Africa), the time frame might be extended to Pliocene.

The decision of choosing the time frame is more complex for areas in between the two rates of activity defined above, because the activity of the area, despite being definable as cratonic, can be influenced by near active tectonic structures defined above. For these cases, one approach is the estimation of the deformed area around the major faults identified at the regional scale at the beginning of the seismic hazard assessment project. A preliminary regional geological, geophysical, geotechnical and seismological (GGG-S) database would be useful to calibrate the time frame for fault capability assessment, assuming that these major structures can induce significant deformation on surrounding areas during major earthquakes and significant stress can be transferred to surrounding faults. Longer time frames may be used when the site is far away from the potentially deformed areas. An example of the deformed area using the InSAR data is given in Fig. A–1.

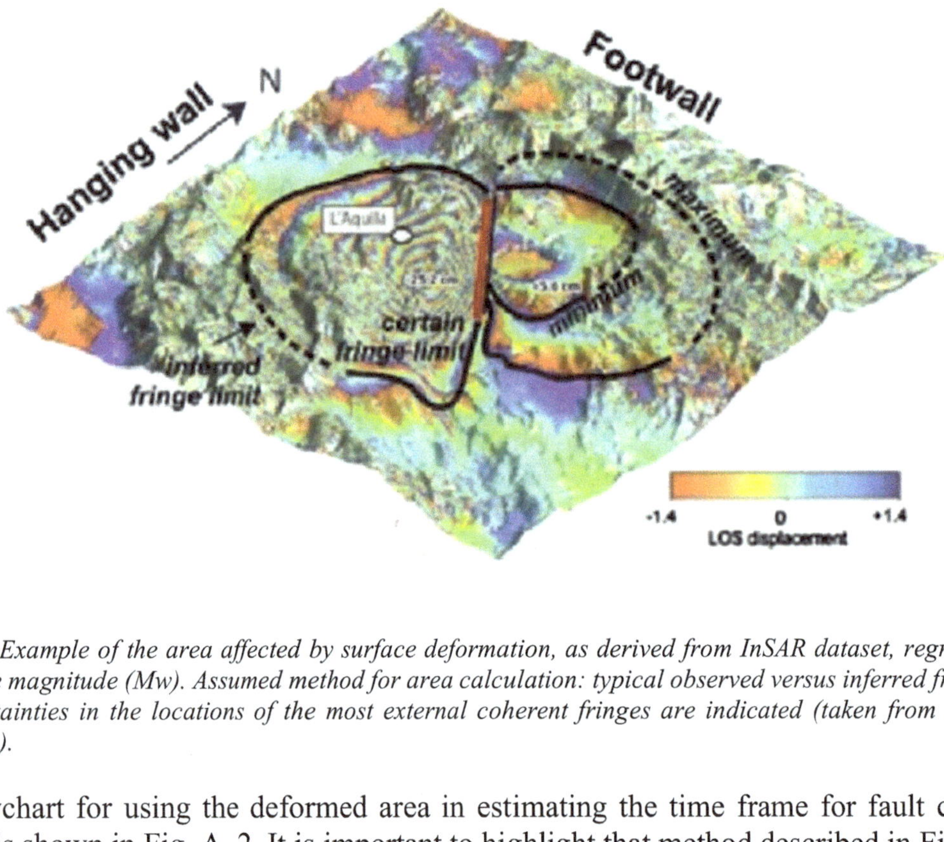

FIG. A–1. Example of the area affected by surface deformation, as derived from InSAR dataset, regressed over earthquake magnitude (Mw). Assumed method for area calculation: typical observed versus inferred fringe limits and uncertainties in the locations of the most external coherent fringes are indicated (taken from [A-2] with permission).

The flowchart for using the deformed area in estimating the time frame for fault capability analysis is shown in Fig. A–2. It is important to highlight that method described in Fig. A–2 is more strictly of interest for States that are still in the process of establishing the time frame for fault capability assessment. However, it might also be of interest for other States that have already defined such a time frame based on rather heterogeneous criteria.

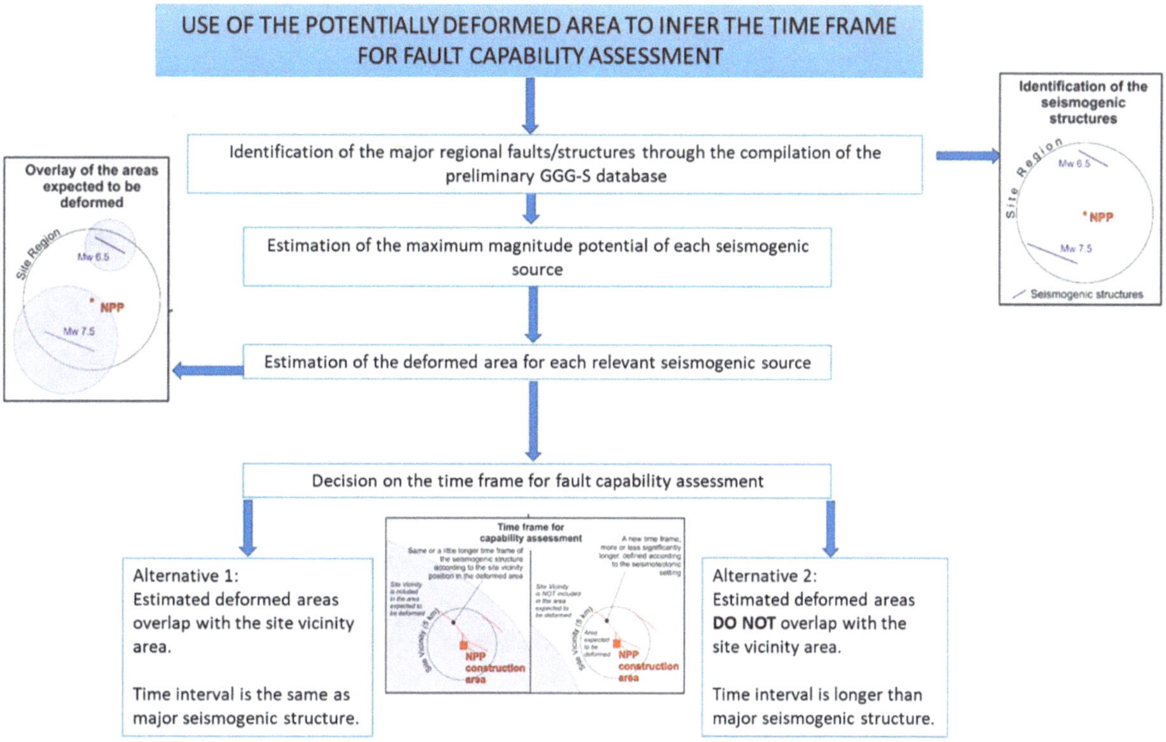

FIG. A–2. Proposed workflow for the use of the potentially deformed area to infer the time frame for fault capability assessment within site vicinity of nuclear installations.

Based on the potential maximum magnitude of the fault, the size of the area that may be significantly deformed by co-seismic slip may be estimated for each seismogenic structure on the basis of a preliminary regional GGG-S database including the major regional faults/structures derived from literature data, using the empirical relationship of Fig. A–3, in which the deformed area is directly proportional to the magnitude of the earthquake.

The estimated deformation areas may or may not overlap with the site vicinity area as shown in Alternatives 1 and 2 in Fig. A–2, respectively. If a fault in the site vicinity lies within the estimated deformed area, the same (or a little longer) time frame for the major seismogenic structure may be applied to assess the capability of the fault within the site vicinity (Alternative 1). If the fault within the site vicinity is outside the estimated deformed area, a new time interval that is significantly longer than the major seismogenic structure has to be used for this fault (Alternative 2).

Selection of the time frame may also consider isostasy due to loading or unloading (e.g. glacial rebound or forebulge), induced seismicity, and volcanism (e.g. Ref. [A–3]).

Once the time interval for fault capability assessment is defined, a database of information and investigations is needed. Paragraph 3.1 of SSG-9 (Rev. 1) [A-1] states:

> "A comprehensive and integrated database of geological, geophysical, geotechnical and seismological information should be compiled in a coherent form for use in evaluating and resolving issues relating to hazards generated by earthquakes."

This database (i.e. the GGG-S database) will allow the identification of the seismogenic sources and the fault pattern affecting the areas of investigation at different scales. Characteristics and parameters of these faults will be provided in the 'project fault portfolio'.

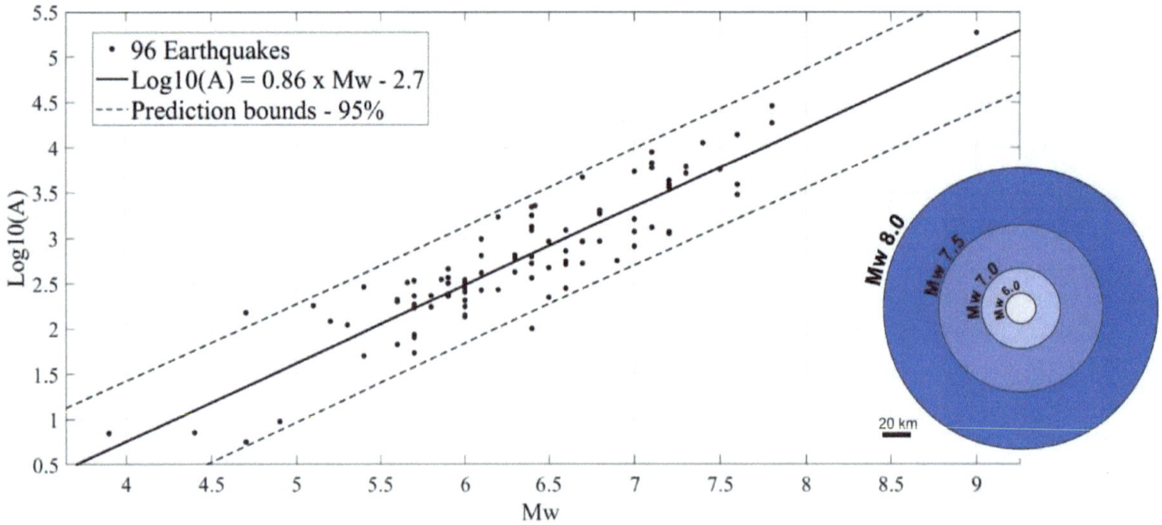

FIG. A–3: Empirical regression of the size of the area (A) affected by significant crustal deformation with the magnitude of the earthquake (M_W) (taken from [A-4] with permission).

REFERENCES TO ANNEX

[A–1] INTERNATIONAL ATOMIC ENERGY AGENCY, Seismic Hazards in Site Evaluation for Nuclear Installations, Specific Safety Guides SSG-9 (Rev. 1), IAEA, Vienna (2022).

[A–2] LIVIO, F., SERVA, L., GÜRPINAR, A., Locating distributed faulting: contributions from InSAR imaging to probabilistic fault displacement hazard analysis (PFDHA), Quaternary International **451** (2017) 223-233, https://doi.org/10.1016/j.quaint.2016.09.034.

[A–3] GÜRPINAR, A., SERVA, L., A risk informed engineering approach to consider faults near an NPP, Transactions. SMiRT-23 Manchester, United Kingdom (2015) Division IV, Paper ID 423.

[A–4] LIVIO, F., FERRARIO, M.F., Empirical Relation among Earthquake Magnitude (Mw) and Deformed Area as Imaged by InSAR, Zenodo (2024), https://doi.org/10.5281/zenodo.11030647.

LIST OF ABBREVIATIONS

AD	average displacement
AFOE	annual frequency of exceedance
FDHI	Fault Displacement Hazard Initiative
FW	foot wall
GGG-S	geological, geophysical, geotechnical and seismological
HCLPF	high confidence low probability of failure
HW	hanging wall
MD	maximum displacement
M_{min}	minimum magnitude
M_W	moment magnitude
N	normal (mechanism, earthquake, or style of faulting)
NPP	nuclear power plant
PFDHA	probabilistic fault displacement hazard analysis
PSA	probabilistic safety assessment
PSHA	probabilistic seismic hazard analysis
R	reverse (mechanism, earthquake, or style of faulting)
SR	surface rupture
SS	strike-slip (mechanism, earthquake, or style of faulting)

CONTRIBUTORS TO DRAFTING AND REVIEW

Abrahamson, N.	University of California, Berkeley, United States of America
Annaka, T.	TEPSCO, Japan
Anooshehpoor R.	US Nuclear Regulatory Commission, United States of America
Baize, S.	Institute for Radiological Protection and Nuclear Safety, France
Baumont, D.	SEISTER SAS, France
Bonadeo, L.	Consultant, Italy
Boncio, P.	University of Chieti-Pescara, Italy
Bozorgnia, Y.	University of California, Los Angeles, United States of America
Chen, R.	California Department of Conservation, United States of America
Chiou, B.	California Department of Conservation, United States of America
Contri, P.	International Atomic Energy Agency
Fernandez, J. A.	Fugro, United States of America
Fukushima, Y.	International Atomic Energy Agency
Gulerce, Z.	International Atomic Energy Agency
Lavrentiadis, G.	University of California, Los Angeles, United States of America
Manchuel, K.	Electricité De France, France
Milliner, C.	Caltech, United States of America
Moss, R.	California Polytechnic, United States of America
Ozacar, A.	Middle East Technical University, Türkiye
Petersen, M.	United States Geological Survey, United States of America
Sakai, T.	Central Research Institute of Electric Power Industry, Japan
Sarmiento, A.	University of California, Los Angeles, United States of America
Serva, L.	Consultant, Italy
Sharon, M.	Geological Survey of Israel, Israel
Thio, H. K.	AECOM, United States of America
Thompson, S.	Lettis Consultants International Inc, United States of America

Valentini, A.	International Atomic Energy Agency
Viallet, E.	Electricité De France, France
Visini, F.	INGV, Italy
Youngs, R.	WOOD, United States of America

Technical Meetings

Vienna, Austria: 2–4 November 2021

Consultants Meetings

Vienna, Austria: 20 January 2021, 14 April 2021, 20–22 June 2022, 30 May–2 June 2023

IAEA
International Atomic Energy Agency

CONTACT IAEA PUBLISHING

Feedback on IAEA publications may be given via the on-line form available at:
www.iaea.org/publications/feedback

This form may also be used to report safety issues or environmental queries concerning IAEA publications.

Alternatively, contact IAEA Publishing:

Publishing Section
International Atomic Energy Agency
Vienna International Centre, PO Box 100, 1400 Vienna, Austria
Telephone: +43 1 2600 22529 or 22530
Email: sales.publications@iaea.org
www.iaea.org/publications

Priced and unpriced IAEA publications may be ordered directly from the IAEA.

ORDERING LOCALLY

Priced IAEA publications may be purchased from regional distributors and from major local booksellers.

25-01494E

FSC
www.fsc.org
MIX
Papier | Fördert gute Waldnutzung
FSC® C083411

Zeitfracht Medien GmbH
Ferdinand-Jühlke-Straße 7
99095 Erfurt, Deutschland
produktsicherheit@kolibri360.de